A PLUME BOOK

SAN FRANCISCO IS BURNING

DENNIS SMITH, a former New York City firefighter, is the founding editor of *Firehouse Magazine* and the bestselling author of eleven books, including *Report from Ground Zero*, *Report from Engine Co. 82*, and *A Song for Mary*. He is currently chairman and CEO of First Responders Financial, a company created exclusively for first responders, and lives in New York City.

"So riveting it is enraging . . . [Smith's] message is the one that matters most."
—*San Francisco Chronicle*

"Riveting." —*The Washington Post*

"A finely woven human story of tragedy, death, heroism and blunder . . . This book is an eye-opener in many ways, and a good read, to boot."
—The Associated Press

"Excellent." —*Star Tribune* (Minneapolis)

"It moves like a shot. . . . Smith narrates all this with command of both close-ups and long shots, and he pays tribute to the heroes." —*Palm Beach Post*

"Vivid." —*Kirkus Reviews*

"While Smith does reveal the factors that made the fire so devastating, such as lack of building codes, flammable materials, and insufficient firefighting infrastructure, he concentrates on the human side, effectively telling the many stories of heroism, stupidity, cowardice, strength, bad luck, and good fortune bred by the fire. This is a readable and exciting book." —*Library Journal*

"Dennis Smith has written an American epic, a masterwork that not only recaptures the lost world of 1906 San Francisco, but also fills page after page with vivid human beings confronting catastrophe. This is a must-read for Americans in the age of terrorism. Smith teaches so much we need to know. Simultaneously his literary skills mesmerize us. Best of all he inspires."
—Thomas Fleming, author of *The Illusion of Victory: America in World War I*

"Dennis Smith has always been the chronicler of catastrophic fire, but he always brings to his narratives a powerful sense of the complexity of the human beings who have to deal with its elemental horror. *San Francisco Is Burning* is worth reading just for its wonderful creation of the rich, venal, exuberant society of 1906. The word 'denizens' seems to have been created for his picture of the people who rule the finances and power blocs of San Francisco on the eve of the city's great earthquake and fire. When earthquake and fire are added, and the impulsive bravery by which the fire is at last quenched, we are left with a wonderful tale which the reader will want to finish at one gulp." —Thomas Keneally

ALSO BY DENNIS SMITH

NONFICTION

Report from Ground Zero: The Story of the Rescue Efforts at the World Trade Center

Report from Engine Co. 82

A Song for Mary

Firehouse

Dennis Smith's History of Firefighting in America

The Aran Islands—a Personal Journey

The Fire Safety Book

Firefighters: Their Lives in Their Own Words

FICTION

The Final Fire

Glitter & Ash

Steely Blue

FOR CHILDREN

The Little Engine That Saved the City

Brassy the Fire Engine

SAN FRANCISCO IS BURNING

The Untold Story of the 1906 Earthquake and Fires

DENNIS SMITH

A PLUME BOOK

PLUME
Published by Penguin Group
Penguin Group (USA) Inc., 375 Hudson Street, New York, New York 10014, U.S.A.
Penguin Group (Canada), 90 Eglinton Avenue East, Suite 700, Toronto, Ontario, Canada M4P 2Y3 (a division of Pearson Penguin Canada Inc.)
Penguin Books Ltd., 80 Strand, London WC2R 0RL, England
Penguin Ireland, 25 St. Stephen's Green, Dublin 2, Ireland (a division of Penguin Books Ltd.)
Penguin Group (Australia), 250 Camberwell Road, Camberwell, Victoria 3124, Australia (a division of Pearson Australia Group Pty. Ltd.)
Penguin Books India Pvt. Ltd., 11 Community Centre, Panchsheel Park, New Delhi – 110 017, India
Penguin Books (NZ), cnr Airborne and Rosedale Roads, Albany, Auckland 1310, New Zealand (a division of Pearson New Zealand Ltd.)
Penguin Books (South Africa) (Pty.) Ltd., 24 Sturdee Avenue, Rosebank, Johannesburg 2196, South Africa

Penguin Books Ltd., Registered Offices: 80 Strand, London WC2R 0RL, England

Published by Plume, a member of Penguin Group (USA) Inc.
Previously published in a Viking edition.

First Plume Printing, September 2006
1 3 5 7 9 10 8 6 4 2

Title page photo courtesy of Historic American Buildings Survey Fireman's Fund Record Vol. XLVI, No. 4, April 18, 1906 HABS, CAL, 38-SANFRA, 105-2, as found in the HABS/HAER/HAL Collection.

REGISTERED TRADEMARK—MARCA REGISTRADA

The Library of Congress has catalogued the Viking edition as follows:
Smith, Dennis.
San Francisco is burning : the untold story of the 1906 earthquake and fires / Dennis Smith.
p. cm.
Includes bibliographical references and index.
ISBN 0-670-03442-8 (hc.)
ISBN 0-452-28759-6 (pbk.)
1. Fires—California—San Francisco—History. 2. San Francisco Earthquake, Calif., 1906. 3. San Francisco (Calif.)—Buildings, structures, etc. 4. Firefighters—California—San Francisco. I. Title.
TH449.S26S65 2005
363.37'09794'61—dc22 2005046112

Printed in the United States of America
Original hardcover design by Nancy Resnick
Map by Jeffrey L. Ward

TO WHOM GRATITUDE IS ABUNDANT

Rick Kot, a generous and caring editor who brought me through the difficult and wrenching days of writing *Report from Ground Zero,* helped me enormously in making sense and story from the voluminous records, always balancing the right amounts of humor and editorial strength. I so appreciate his friendship.

And Clare Ferraro, the president of Viking, again gave me her special brand of enthusiasm that this time propelled me westward with needed confidence.

I owe a special debt to Robert Courland, who has an authority of knowledge that helped me along every path of American, international, and scientific history. His care and concern for finding the most reliable citation and reference is gratifying, and his observations are always memorable and consequential.

Robert Nason, Ph.D., whose vast geoscience reading and expertise were always available to me and who provided many hours of insightful conversation.

Scott Peoples, a retired San Francisco Fire Department assistant chief, whose willingness to take on every request is matched by his respected expertise on the unique fire-related problems of the Bay City.

Each of the following men and women deserves a paragraph or more regarding their friendship and helpfulness in accomplishing the needed research into 1906:

Bill Murray, third in line of a great firefighting family

Andy Casper, former chief of the San Francisco Fire Department

Don Casper, son of a firefighter and friend

John Hanley, president of SFFD Firefighter's Union, and his supportive board

Gladys Hansen, whose keen skill and integrity can by counted on by every researcher into the 1906 San Francisco earthquake and fire. And her son Richard, who has helped Gladys build the most important collection of 1906—the San Francisco Virtual Museum.

Joanne Hayes-White, chief, San Francisco Fire Deparment
Jim Doyle, *San Francisco Chronicle* journalist
Jim Baker
Judith Barringer
Peter Doyle, who lent James D. Phelan's private narrative of the fire
Steve and Florence Goldby
Alessandra Lusardi, whose responsiveness is as reliable as 911
Mike McDowell, collector and bibliophile
George Montgomery, and his son Jonathan, who, along with their friendship, provided important accounts of the Spreckels family history
Kevin Mullin, retired inspector of the San Francisco Police Department, writer, and police department historian
Paulette and Shep Pollack
Adolph Rosekrans
Pete Salenger
Al Sassas
Ann Seymour
Darryl Siry of Fireman's Fund Insurance Company
Kevin Starr, the important California historian
Dave Strohm
Richard Tower
Putney Westerfield
Gregg Young, DDS and SFFD
Rebecca Livingston of the National Archives and Records Administration
Josephine Schallehn
Marie Chan of Wells Fargo Museum
Paul Conroy, a fire commissioner of San Francisco
Captain John Dellinges of the SFFD
Mario Travino, former chief of the SFFD
Inez Cohen, director of the Mechanics' Institute Library
Nonni Delimur
Reg Stocking
Thomas Wright

And in New York, all the usual crowd willing to go to lunch on short notice: Brendan and Elizabeth; Dennis, Jackie, Julia, and Fiona; Sean, Christina, Carlin, Henry, and Fayre; Deirdre; and Aislinn.

To all of our firefighters, police officers, nurses, and EMTs
who put their own lives on the line at every emergency,
terrorist attack, or natural disaster.
Thanks for being there.

The San Francisco Fires

April 18–21, 1906

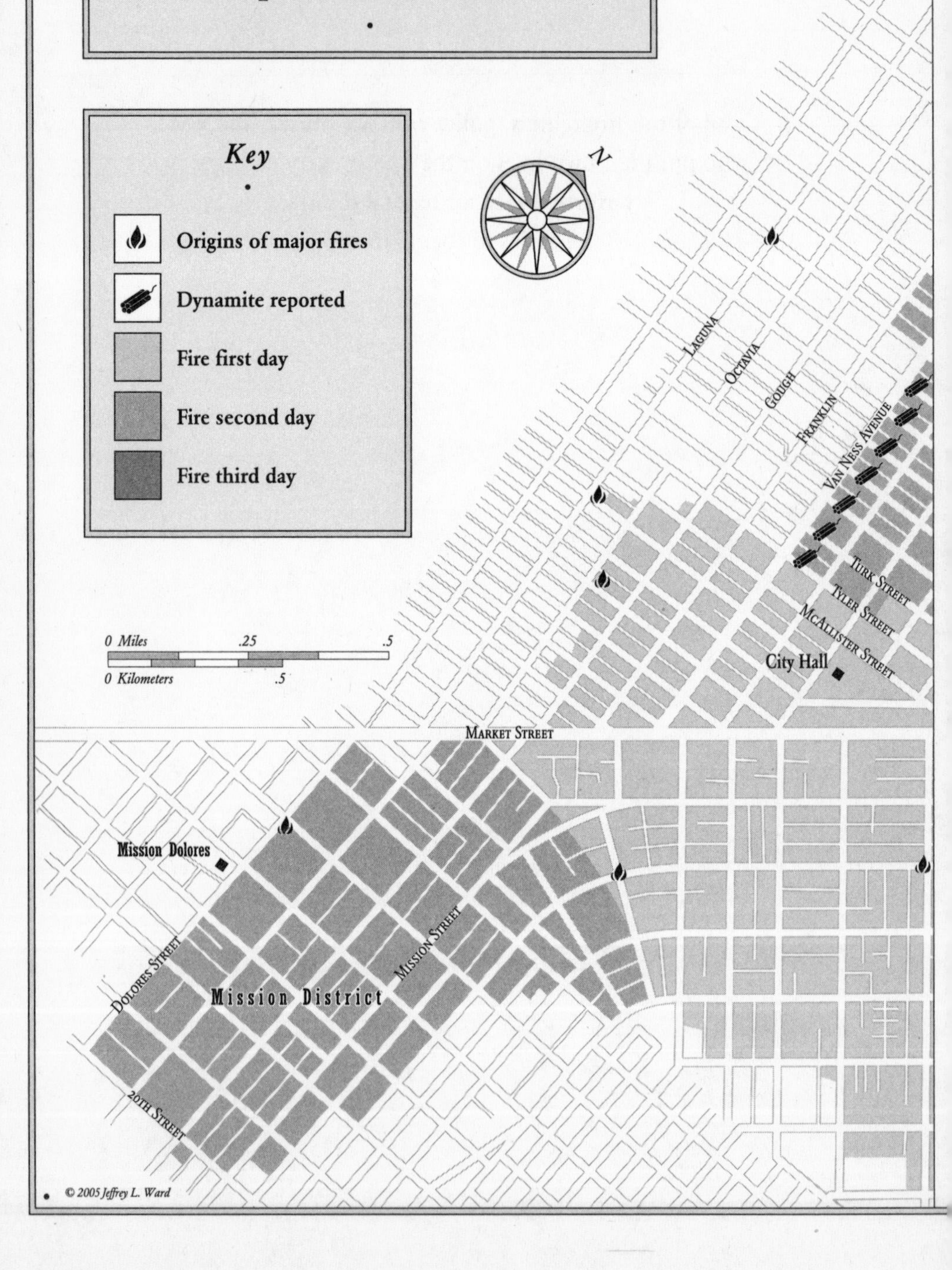

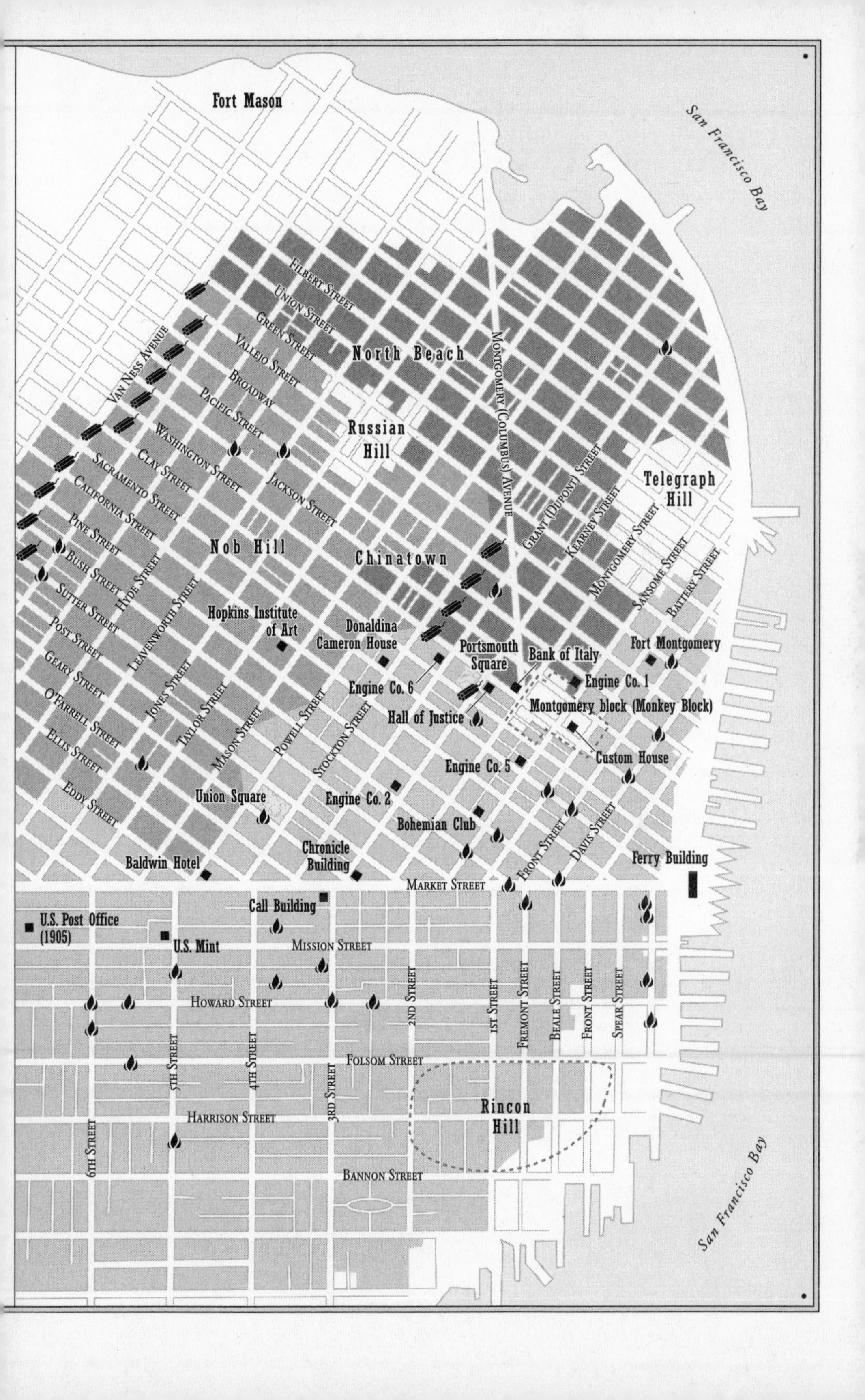

Fort Mason
San Francisco Bay
Filbert Street
Union Street
Green Street
Vallejo Street
Broadway
Pacific Street
Washington Street
Clay Street
Sacramento Street
California Street
Pine Street
Bush Street
Sutter Street
Post Street
Geary Street
O'Farrell Street
Ellis Street
Eddy Street
Van Ness Avenue
Hyde Street
Leavenworth Street
Jones Street
Taylor Street
Mason Street
Powell Street
Stockton Street
Jackson Street
North Beach
Russian Hill
Nob Hill
Chinatown
Telegraph Hill
Montgomery (Columbus) Avenue
Grant (Dupont) Street
Kearney Street
Montgomery Street
Sansome Street
Battery Street
Hopkins Institute of Art
Donaldina Cameron House
Portsmouth Square
Bank of Italy
Fort Montgomery
Engine Co. 6
Engine Co. 1
Montgomery block (Monkey Block)
Hall of Justice
Custom House
Engine Co. 5
Engine Co. 2
Union Square
Bohemian Club
Front Street
Davis Street
Chronicle Building
Baldwin Hotel
Ferry Building
Market Street
Call Building
U.S. Post Office (1905)
U.S. Mint
Mission Street
Howard Street
Folsom Street
Harrison Street
Bannon Street
6th Street
5th Street
4th Street
3rd Street
2nd Street
1st Street
Fremont Street
Beale Street
Front Street
Spear Street
Rincon Hill
San Francisco Bay

INTRODUCTION

I am standing on the bow of the *Guardian,* one of the San Francisco Fire Department's two fireboats, enjoying the excitement of the moment. I am here, at last, on the waters of San Francisco Bay, in the shadow of imposingly tall buildings and verdant rolling hills.

The story of San Francisco's great earthquake and fire of 1906 is one that inspired me some years ago when I read the first of what would be scores of books and hundreds of miscellaneous articles about the times. Even in the most cursory reading of those pieces I could see that the firemen of San Francisco performed not only heroically but also historically. Yet there are no satisfactory accounts of the men who confronted one of the greatest conflagrations our nation has known. Why hadn't this story of a tremendously difficult, block-to-block four-day firefight been recorded, and why hadn't the firemen of San Francisco been given their due?

At the same time, I began to wonder about a naval lieutenant who had been cited as making a brave stand against the fire on the waterfront. Why hadn't his photo been included in the montage of the more than 450 photos published in 1907 under the title *Heroes of the Great Calamity*? And in all the books and accounts, why was there no background information about this man whose contribution had clearly been critical?

Instead, the stories told about 1906 have mostly centered on the corruption of the times, on rumors of wanton murder, on the pervasive lack of water, and on the terrible ruination. Surely, I asked myself, within a disaster that seemed to be mismanaged and out of control and in a municipal environment of mostly fools and criminals, there had to be some honor somewhere. The answer would take an entire book.

Every writer must ask why he should devote an enormous amount of time to a particular subject—what is it about that subject that is compelling enough to justify years, sometimes decades, of research and writing. Given my interest in the history of firefighting, there are many important catastrophes I could have written about, fires that thwarted and then inspired the growth of cities in America—New York in 1835, Chicago in 1871, and Baltimore in 1904. There are lesser known but equally extraordinary fires, like the Peshtigo fire of 1871, which killed two thousand, or the *General Slocum* fire of 1904, in which more than a thousand perished, most of them children. I know these stories well, and they have all interested me in the way they helped define and shape our culture. Except for one book written by a San Francisco librarian and a fire chief, the generosity of facts laced with the threads of rumor and fiction in most of the accounts I have read serve only to confuse the story. In fact, the fire of 1906 challenges most of the conclusions that have been officially made regarding it.

Perhaps the story has remained incomplete because it is so large, or because its historians have overreached for dramatic effect, or because so many documents from the fire department and from the city's own Committee on History have vanished. Or, perhaps, there was a deliberate subsequent effort by city officials and businessmen to diminish the tragic impact on the lives of hundreds of thousands of people in an attempt to diminish the public relations damage of a cataclysm.

Finally and inescapably what interested me most about San Francisco's fire is that it was caused by an earthquake, an event that no one in the emergency services can predict, or, since we can never know the magnitude of a particular quake's power, adequately prepare for. Because the possibility of earthquake damage in today's overbuilt and overpopulated cities all over the world continues to be a compelling and profoundly important subject, I could not think of a more meaningful way for me to spend my time as a reader or as a writer.

As the boat cuts through the waters fronting the Bay City, one of the most famous sites in the world, I begin to talk with Scott Peoples, a friend who has recently retired after twenty-eight years of service with the San Francisco Fire Department, the last thirteen as a fire chief. A former assistant chief of department, he is one of the most respected firefighters I have met in America. The boat is idling as it faces south toward the city, about a mile out from the marina district, and we are dipping with the mildly choppy waters. It is an easy, still day. On my right, the Golden Gate Bridge radiates through the poetry of its red harp strings a glowing harmony to the horizon. Beyond the shoreline beneath the bridge, and beyond the rows of multimillion-dollar hill homes, are the powerful Pacific, Hawaii, and China. On my left, the Oakland Bay Bridge to San Francisco flexes its blue-steel strength and slender rhythm as it looms in two parts, each like an elongated flying buttress, from Oakland and from the island of Yerba Buena.

"The San Francisco fire was bigger than the one in London," I remind Scott over the whine of the engines. "Bigger than Moscow, Rome, Lisbon, New York, Boston, and Chicago. In fact, outside of war, the San Francisco fire of 1906 is bigger than any metropolitan fire in history." The four-day event took more than 3,000 lives, burned through 28,188 buildings, flattened 522 blocks, destroyed tens of churches, 9 libraries, 37 national banks, the Pacific Stock Exchange, 3 major newspaper buildings (the *Call, Chronicle,* and *Bulletin*), 2 opera houses, and the largest, most richly appointed imperial hotel in the era of turn-of-the-last-century opulence. More than 200,000 people were burned out of their homes—men, women, and children who found themselves wandering smoke-filled streets with no claim to a future except that they were alive and that they owned the clothes on their backs.

Today, on this serene spring morning, there is no evidence of tragedy apparent, but beneath this restful tranquillity is an acknowledged and accepted understanding that the unpredictable can occur again, and destroy, as it has several times in its recorded past.

The story of the San Francisco earthquake and fires is a hard story of numbing imbalance, of corruption and virtue, of stupidity and enlightenment, and most of all, of cowardice and courage. It is also a story of peculiar fire-loading conditions and of a geologic vulnerability that is very dangerous—factors that have proven to be ruinous.

I have worked to place the event in the context of its scientific, its social, legal, and military history, and within the extraordinary cultural changes that overlook America as the nineteenth century began to confront the challenges of the twentieth, as it worked its way through the gilded age and into an age of political reform at home and muscle flexing abroad—an age that would take us innocently into something we would call a world war.

San Francisco, a city of enduring culture and taste, has been known throughout the world as the Paris of the Pacific. But its many contradictions have occasionally led to its being known as the Baghdad by the Bay—a city of historic wickedness and ferocity, with more than a few vestiges of the frontier in its character. In 1906, it had the highest per capita murder rate in America and was the only city in the country where a killing in a boxing ring had no legal repercussion. While it had a church for every 2,500 citizens it also had a saloon for every 250 of them. At the same time it was inhabited by workers, clerks, entrepreneurs, and established bankers and builders—men and women who had conquered the hills and were building an empire that had already made the city by the bay the leading metropolis west of the Mississippi.

The sky, with its soft cumulous designs, is as vivid as a Vermeer, and we try to imagine this polished blue heaven before us turning suddenly black from the smoke of hundreds of fires. Within the black, vague and erratic, a yellow glow jumping and skipping though a whirling darkness. This is what Lieutenant Frederick Freeman, USN, would have seen that terrible morning in April 1906. He must have known that the fire would take the city. Scott agrees. The smoke clouds he watched from the bridge of the USS *Preble* were too wide and furious to be fought by one or even a dozen fire departments. This is Lieutenant Freeman's story, and it is the story of one man, Rudolph Spreckels, who put his conscience before the social well-being of his family and his friends. It is also the story of the strong and enduring firefighters of Engine 1 and all the engine and truck companies of the San Francisco Fire Department. It is, finally, the story of the fearsome and devastating consequences of the sudden and unexpected movement of the earth—the side shifting of a huge mass of the earth's crust—upon a major city.

For now, that city seems to convey only the calm, the assurance, the wealth and certainty that have come from nearly one hundred years of commercial progress, a rich growth that has been built from the ashes of 1906 and the many fires before it.

As Chief Peoples knows well, after almost three decades of firefighting, fires are always ultimately contained and that some phoenix will always rise from the ashes. But, as the chief reminds, it is not the possibility of fire that is San Francisco's nemesis.

It is earthquakes. And earthquakes are as certain as the tides.

CHAPTER 1

John Pond had the letter in his hand as he knocked softly on the door. Hearing no answer, he entered the room quietly and saw his old friend sleeping. Now frail and having the look of weariness that comes with a beaten body, Frederick Freeman was perfectly still, the thin bedspread pulled up to his shoulders.

He seemed so distant now from "Frisky," the nickname he had always been called by the officers and, out of earshot, by the enlisted men as well. It was an endearing name, one that agreeably conveyed his personality. He had been a tall man, thin and straight, a commander through and through. That his command had loved him is certain. He was always the first on his feet to rush headlong into the unrelenting challenges, big or small, presented by the sea, always urging them in that easy, authoritative style of his, "Okay, men, let's sock it to 'em!"

Self-assured and confident men can oftentimes seem arrogant, but few officers in the navy ever had more consideration and gracefulness, mixed with equal parts of a kind of determined certainty. The seamen always followed his lead for he was a sailor first, then a leader.

But today, John Pond was thinking that Frisky, with his deep blue eyes closed and his head barely resting against a hard pillow as if sculpted for a ceremonial sarcophagus, seemed to be caught in a sort of lethargic remove, his demeanor so unlike Pond's memories of him.

It was a sunny morning in late January 1941, and Pond had traveled to Soledad, a one-horse farming town 150 miles from San Francisco, to tell Frisky the news. Freeman had come here to the Pasquale Hotel because it was cheap, just barely affordable on his meager income. He had a small, well-ordered room, but it was dark and musty. Still, he needed to rest because he was unwell. He knew he had cancer, and that it was bad.

Pond considered waking his old friend to read the letter aloud to him, if only to bring some change, something that might sparkle within the dusty and moribund room. But Freeman appeared so peaceful. If not for the lulled rising and lowering of the sixty-six-year-old man's chest, Pond might have thought him dead.

John Pond, himself a recently retired naval commander, had been quite surprised just a few months earlier when he had come upon his former shipmate walking up the narrow main street of Burlingame, California. Burlingame was an established and proud community, many of whose well-off residents were the heirs of San Francisco's old railroad and mining wealth or magnates of the newly thriving real estate and investment businesses. Against such a well-manicured, colorful background the former captain's skin looked especially gray and sallow. His once-bulging chest was now caved and reduced and whatever assurance he once held in his eyes had been replaced by a startling sadness. Yet his head was high, and his shoulders remained erect, as if to announce that however broken and down and out he might be, he was nevertheless, at least in his heart, a commanding officer in the U.S. Navy.

A brown-corduroy-covered easy chair stood next to the bed, and Pond took a seat in it, sliding the letter carefully into the inside pocket of his jacket. He would wait until his skipper awakened.

It seemed as if the easy chair embraced him with an old-friend warmth, and in its comfort John Pond studied the figure before him. He began slowly to recall what they had gone through together, to consider the changing fortunes of Frederick Freeman. He had led the life of an esteemed man, one who had in 1906 reached the height of triumph. Nearly forty years later he was crushed, having fallen to the depths of ruination.

How difficult it must have been for so solid an individual to have borne the burdens of his past, Pond thought, his eyes never wavering from the profile lying before him. But he did bear them, stoically and with acceptance, living a reclusive life through twenty-five years of disgrace and ostracism. And, now, in this little hotel room where he had come to be alone and to repair, he was still bearing them with whatever strength remained in his heart, if not in his body. He did not know that the country might yet remember those four tragic days in San Francisco, when America was still young and its men and women were ready to give their lives to make it what it could be.

CHAPTER 2

On Tuesday night, April 17, 1906, Lieutenant Frederick Newton Freeman, USN, commander of the torpedo boat destroyer USS *Perry,* took his place at the green steel table of the officers' mess, a space not much larger than any of the four officers' berths on the ship. He was a handsome man, and some would say he was made handsomer by a thick and prominent nose, which he kept held high.

He was seated with Midshipman John Pond, who at twenty-two was a man just nine years younger than his skipper. Pond was an impressive officer, sharp, well built, and keenly intelligent, and Freeman was undoubtedly happy to have him on board.

The *Perry* and her sister ship, the *Preble,* had been on maneuvers testing the newly developed wireless telegraphy system within the Pacific Fleet. They had been off Magdalena Bay in Mexico, just south of the California border the previous week, when Freeman received orders from Admiral Caspar Goodrich, the commander of the USS *Chicago,* the fleet's flagship, to head north to the Mare Island Naval Base near San Francisco to meet their maintenance schedules, for both boats were due for annual overhauls. Being in dry dock was an opportunity for the *Preble*'s commanding officer and crew to go on extended leave, but Freeman was to stay on board his boat with his two junior officers, Pond and Ensign Wallace Bertholf. The *Preble*'s overhaul had just been completed, and the *Perry* now had its motors on blocks.

Freeman had yearned for a career at sea since boyhood, and in 1891, at sixteen, he entered the service and began to work himself up through the ranks. He went to Annapolis and did well. Though there were many officers who predicted an admiralty for Frisky, there was an equal number who were determined that he rise slowly and less securely. Freeman would come to see that there were unclear but certain divisions of support and loyalty between naval officers that could unsettle a career in the differing commands of the navy.

When the Spanish-American War broke out in 1898, he looked forward to serving in the Philippines or even being involved in the Cuban action, but the Atlantic and Asian fleets won all of the early battles, and the war was soon over. Still, Freeman was dutiful, and his career blossomed. Now, at thirty-one, having spent almost half his life in the navy, he was a lieutenant, and in the language of the old navy he was "master and commander" of the torpedo boat destroyer *Perry*. He was addressed formally as Captain Freeman, but on shipboard the men called him "skipper."

Freeman's work was exceedingly important in naval defense strategies, for the advent of small, fast torpedo boats in the late nineteenth century was the bane of admirals. These inexpensive torpedo boats could easily outmaneuver and outrun, and so easily sink, the hugely expensive capital battleships and frigates. To fight them, a new kind of ship was developed, a truncated destroyer, low to the water, thin and fast. The *Perry* and the *Preble* were of this Bainbridge design, and, in 1899, both ships were laid down within two days of each other, built by the Union Iron Works in San Francisco Bay. The *Perry* was 250 feet long and only 23 feet wide. This sleek style, with its four tilted smokestacks pumping the power from four engines, made her one of the fastest boats in the navy—indeed, in 1905, Lieutenant Freeman was awarded the Pacific Fleet's Torpedo Boat Destroyer of the Year trophy. The *Perry* could easily do 29 knots, 10 knots faster than most cruisers. Freeman was proud of his boat, and he was always ready for action. He knew that because of the narrow beam and low profile, his was a hard boat to target, and so when the time came, he would be able to get close to the capital ships and bring them down with his short-range torpedoes.

For the moment, though, the *Perry* was a crippled ship, and Freeman and Pond were taking pleasure in a dinner that was being served while in the still port waters. Pond had been aboard for just fifteen days, and he and Freeman were, little by little, getting to know each other.

"Tell me about your father," Lieutenant Freeman said. It was a question that made the younger man uncomfortable, for like most sons of personages he did not want to rely on his father's reputation to make his way. But Freeman could see that at most there would be a reluctant response and so quickly redirected the question. "Never you mind," he said. "I guess I know enough about your father anyway—enough to respect him much."

A sensitive perception, young Pond thought, relieved that he did not have to answer. His father, Admiral Charles Fremont Pond, was well known then for his leadership of many sea battles in the Philippines, but would come to be remembered primarily for being one of the fathers of naval aviation (while captain of the USS *Pennsylvania* in 1911, he built the first landing strip on a navy vessel) and for being the man to suggest and then to build the naval base at Pearl Harbor.

John Enoch Pond resembled his father in many respects. In 1905, he became the first graduate of the U.S. Naval Academy from the territory of Hawaii. He was as consciously loyal to the navy as he was to his father. The Ponds were a naval family, and it went without saying that if he was ever to have sons, they would be sailors, too. Young John, responsible for the ship's stores and ammunition, was alert, innovative, and fully involved in his work. And he had a sense of humor.

Lieutenant Freeman was born in Fort Wayne, Indiana, in 1875, and at thirty-one he had maintained through his travels a midwesterner's view of the world. Seriousness and willingness were the two greatest qualities a man could possess, and humor was the crown of character.

"I would rather sing the role of Don José," Lieutenant Pond now added, "than tell stories about my family."

Freeman laughed at this reference to an earlier discussion about the San Francisco debut of Enrico Caruso this very night, in *Carmen,* at the Tivoli Opera House.

It was an unusually warm evening, even for an approaching spring, balmy, and windless. A porthole was open, and they could hardly hear the ripples lapping against the pilings.

CHAPTER 3

At about the same time, in the master bedroom of 1900 Pacific Street, one of the most lavish Victorian mansions in the wealthiest of the city's neighborhoods, Rudolph Spreckels sat on the side of the bed next to his wife, Eleanor. The thirty-four-year-old multimillionaire, in white tie and tails, was holding her hand, making certain that she would be all right at home for the evening. She assured him that she would be as perfectly fine as he would be at *Carmen*.

Rudolph kissed her lightly on the cheek. She was a strong and beautiful woman, his Nell, and he was delighted to see how well, at seven and a half months with child, she was doing. She was six years older than Rudolph, and so it was vitally important that during this pregnancy she remain healthy, not only for the child's sake but for her own as well. When they had wed in 1895, the local newspapers had referred to Eleanor J. Jolliffe as one of the most beautiful women in San Francisco, but now the marriage had brought her burdens that she could never have imagined. The pain of losing their little boy, four-year-old Rudolph, five years earlier was still strong with the couple, and Rudolph hoped the new baby would revive Nell's spirits. To lose a firstborn is particularly hard on a mother, and not even the subsequent births of Howard or Eleanor was able to pale the memory of it.

Rudolph began to take his leave when his wife held fast to his hand. She looked into his bright, almost glowing blue eyes, the same blue eyes that had captured her heart eleven years before when she had agreed to marry him, and she kissed him again.

Theirs was a formal relationship, as befitting a household where footmen in white gloves stood at attention behind the chairs at meals. San Francisco, like New York and St. Louis, took its social decorum seriously, and at the turn of the twentieth century if one had the money to live like the princes of Europe, then so elaborate a manner of living became more an obligation than a fancy. San Francisco, after all, was not only the largest city in the West, but it was also the Western center of culture and finance, and its leading citizens were the arbiters of Western style and taste.

The essence if not the refinement of that style was reflected in the huge baroque, Victorian, Italianate, and mansarded homes that were constructed on the top of Nob Hill and along Van Ness Avenue by people like Rudolph, by James Flood of the Virginia City Comstock Lode, and by the railroad barons Charles Crocker, Mark Hopkins, Leland Stanford, and Collis Huntington. Some doubtless went past the boundaries of good taste. There were dozens of these westernized castles, and in them were found Turkish bedrooms of ivory and marble, receiving room ceilings from French *hôtels particuliers,* oak dining rooms from English manors, foyers from the palazzi of doges and cardinals, and horoscope rooms laid out in gemstones—all set in architectural detailing that could only be described as part palace, part *schloss,* part château, and part hostel. They were encased in turrets, dormers, battlements, widows' walks, and mansard roofing, detailed by columns and pilasters topped by ribbon-draped capitols—Doric, Ionic, and full-feathered Corinthians—all bordering swinging garlands; dentil, egg, and dart moldings; and family crests—the painted ladies of a seaside city, the bright kissing cousins of the homes of Bath.

Of the home of one of the big four, the journalist Ambrose Bierce said, "There are uglier buildings in America than the Crocker home on Nob Hill, but they were built with public money for a public purpose."

These showplaces were filled with furniture, artifacts, pictures, and sculptures brought home from European jaunts—one mogul who was considering a painting said to belong to the Medici demanded before closing the deal to first speak with Mr. Medici himself. Conspicuous consumption was a virtue, and because of it the city developed a reputation for ostentation that was spread internationally. "San Francisco is a mad city," Rudyard Kipling wrote, "inhabited for the most part by perfectly insane people whose women are of a remarkable beauty."

On bidding farewell to Nell, Rudolph Spreckels stepped into his carriage and headed down Van Ness. He passed his father's, Claus Spreckels's, home at Sacramento and Clay, just four blocks away, one of the most spectacular buildings in all of America, and he wondered if his parents would appear at the opera that night. Probably not, for they lived quietly. But, still, it was *Carmen,* and a new *Carmen,* it was said, would attract five times the attention of a new *Hamlet.* Rudolph felt a tinge of sadness, as he always did while passing the turreted brownstone mansion, for although they had recently reconciled, he had been estranged because of a business disagreement from Claus, the nation's sugar king, for more years than he cared to remember. Although he had kept in continual touch with the love and kindness of his mother, he profoundly missed his father's sage companionship.

The driver turned down Market Street, passing the large and flamboyant office building of Rudolph's friend James D. Phelan, and then on 3rd Street the famous Spreckels Building, the tallest building west of the

Mississippi, where the Spreckels family's newspaper, the *San Francisco Call,* was published. Then they turned east and headed for the Tivoli Opera House on Mission Street.

Three thousand of the most established members of San Francisco society had made their way to the sold-out house, having paid sixty-six dollars for a box and seven dollars for a balcony seat. It was reported that the higher the prices at the opera the lower the necklines, and brazen décolletage could be seen everywhere.

On the street, Rudolph saw the de Youngs of the *San Francisco Chronicle* and nodded politely—the families had not actually spoken in twenty-two years, when Adolph Spreckels, the second of Claus's four surviving sons, shot Michael de Young at his editorial desk for having accused Claus of defrauding his stockholders in the Commercial and Sugar Company of Hawaii. Fortunately, Michael lived, and Adolph was acquitted on the grounds of reasonable cause, a defense often acceptable in San Francisco history. "Hatred of de Young," observed Ambrose Bierce, "is the first and best test of a gentleman."

The *Call,* bought by Claus in 1895 and published by his first son, John D. Spreckels, since 1897, represented the interests of the progressive wing of the Republican Party while the de Youngs' *Chronicle* was decidedly old-line conservative Republican. The only thing the two papers seemed to have in common was an antipathy for the financial and political arrogance of the Southern Pacific Railroad and its eastern leader, Edward H. Harriman, whose influence was caricatured in their political cartoons as an octopus.

Entering the lobby of the opera house, Rudolph waved to Fremont Older, editor of the *San Francisco Evening Bulletin*. The two men had been seeing much of each other in recent weeks, and indeed had been together just that very afternoon to hear the well-known prosecutor Francis J. Heney report that their first efforts of investigating city corruption were paying off. Rudolph smiled at William and Mrs. Tevis, whose three-inch diamond dog collar, glittering among a sea of tiaras, was the ne plus ultra of extravagant adornment at this toniest event of the year. Will was the son of Lloyd Tevis, who had replaced William G. Fargo as president of the Wells-Fargo Company before moving the bank from New York to San Francisco. Will Tevis had been left part of every gold and silver mine in the West, and more than three hundred thousand acres of prime California real estate. He was now president of the Tevis Land Company, which grew out of a cattle, land, and mining partnership with Ben Ali Haggin, another millionaire land tycoon. Tevis was also a foe of James D. Phelan, the city's former mayor. Both wanted to bring water to San Francisco, but Tevis alone planned to personally profit from the venture and profit at the outrageous level of ten million dollars.

Conspicuously absent from the audience was the current mayor, Eugene Schmitz, who, had he been in attendance, would surely have been in the lobby extending his corrupt availability to the leading businessmen. Described as "the smallest man mentally and the meanest man morally" ever to sit in the mayor's seat, Schmitz had presented himself in the election of 1901 as "a man of the people." Apart from the fact that he had been president of the Musician's Union, there was no evidence that he was particularly concerned about the affairs of workingmen and -women. That he had greed flowing through his German veins and corruption behind his Irish eyes was unquestioned. Rudolph cringed every time he thought of the mayor, a political puppet whose venery was pulling the reputation of his beautiful city into the gutter.

Rudolph saw Phelan sitting with his sister and waved to them grandly. Phelan was a true patriarch among the leaders gathered there that night. He had been elected to the mayor's office three times on the Democratic ticket, but had been forced to retire in 1901 because he was unable to settle a teamsters' strike that had left two hundred boats sitting idly in the harbor, waiting to be unloaded. The strike created great social unrest and financial duress in that the city's Manufacturers' and Employers' Association was determined to break the union and imported armed strike breakers. Sixty percent of the city's business came to a standstill in a general strike, and Phelan was accused of using the police to try to break it. Though the police were in fact used simply to keep order, their very presence at strike rallies proved the mayor's undoing. The strike also presented an ideal opportunity to entrench the new Labor Union Party, a political boss Abraham Ruef, and a theatrical musical conductor named Eugene Schmitz as a potential mayor.

The bachelor heir to a considerable fortune and a graduate of the Hastings College of Law, Phelan was a model of late-nineteenth-century education, propriety, and intelligence. He wrote poetry, was a proponent of the country's City Beautiful architectural movement, and, along with

Rudolph, the prime mover of an investigation to expose the corruptions of the mayor and the board of supervisors. Like many wealthy men in the city at that time, he was also involved in a multitude of investment schemes, many with his friend Rudolph. They had begun their partnership by pooling their money to buy a hotel, and they then developed a bank, a real estate investment company, and a gas and electric company. And just a day earlier, on April 16, 1906, they had consolidated their interests to develop a railroad streetcar system for the city whose power lines would run below the ground. It was a venture in which they stood to make not a penny, for they had promised that if it proved successful, they would sell it to San Francisco for the cost alone. It was also a canny public-interest effort to preserve the beauty of the city.

When he was mayor, Phelan had also bought up much property in his own name, in the Hetch Hetchy Valley 160 miles northeast of the city, and in corridors between there and San Francisco. It was his intention to ensure the development of a reservoir to transfer the land to the city at cost, and he undertook the purchases secretly in order to keep speculators away from what he wanted to be a government-sponsored building plan. Phelan would have been deeply troubled had he been aware that Abraham Ruef, Mayor Schmitz, and the board of supervisors had already been guaranteed one million dollars in kickbacks if Will Tevis's competing Bay Cities Water Company proposal was accepted, an amount that would have a twenty-fold value in twenty-first-century dollars.

Walking toward his seat, Rudolph greeted Mary Leary Flood, a famously genteel woman, who with her husband, James, one of the big four silver kings, was a former bartender. It was rumored maliciously that the family had sold its cow in Ireland to get her a ticket to San Francisco society. Also in the audience were Frederick Kohls, a multimillionaire capitalist who had once been shot by his French maid; Jeremiah Dinan, Mayor Schmitz's police chief, who had been a police corporal until the mayor jumped him over seven captains to head the police department; Rudolph's brother Adolph, from whom, along with his older brother, John D., Rudolph was still estranged. Claus had raised his sons to be of independent mind, and it took just one business misunderstanding within the family trusts to lead them to acrimony and intransigence, Rudolph siding with his brother Gus, in whom he had placed his confidence.

He searched the boxes and around the ornately gilded auditorium with its huge Venetian chandelier, said to be unique in the entire world, but did not see his parents. Perhaps it was better, he thought; their presence might divert him from the pleasure of the music. And the music, one critic promised, would make San Francisco forget its diamonds—an accurate prediction that was borne out as curtain call after curtain call was demanded by the exuberant San Franciscans. Rudolph laughed, thinking that someone was sure to say that because of Caruso's performance the opera should henceforth go by the name *Don José* instead of *Carmen*.

When he left the opera that evening, he did not notice Chief of Department Dennis Sullivan's buggy clanging its way down Bush Street. The fire chief was pulling on the bell string while the operator, his driver, cracked a whip over the head of the tall gelding called Brownie. Sullivan, who had been born in Boston in November 1854, had been the fire chief for exactly thirteen years and thirteen days, and looked ten years younger than his fifty-two years. Passersby stopped in the streets to wave him on, but the chief did not notice, so focused was he on the alarm location. The chief of the fire department responded only to major fires, second alarm or more. There had been forty-eight of them the year before and already that year there had been a few dozen, and one in which two of his best firemen had been killed.

The streets were crowded with first nighters, walking toward Nob Hill or in their carriages on the way to the hotels where they would revel until the early hours, or to the "French" restaurants, like the Poodle Dog and Delmonico's, where food was taken on the first floor, drinks and suggestions received in discreet parlors on the second, and the fulfillment of secret desires in private rooms on the third.

John Barrymore, fresh from his success in *The Dictator,* was on his way to Ben Ali Haggin's mansion at the top of Nob Hill, having been invited to dinner and presumably to see the famous blocklong Haggin stables, said to be more beautiful than most mansions. Barrymore was scheduled to leave the following morning for Australia on one of John D. Spreckels's Oceanic steamship liners. And Elsa Maxwell, well known but not yet famous, was in a carriage on her way to a postopera supper to offer her memorable flair and spirit.

Rudolph, however, had no plans to socialize and returned home, quietly entering so as not to disturb the staff. He undressed and slipped easily into his majestic-size bed. Nell did not stir, and it pleased him to see her so restful.

It had not been easy for Rudolph to find sleep since the recent incorporation of the Municipal Street Railways of San Francisco. Concerns about the required public approval of his underground wire proposal weighed heavily on his mind, especially since he was so aware of the friendship between the money-hungry politicians and Patrick Calhoun, owner of the rival streetcar line, United Railroads. Rudolph knew that he had some kind of arrangement with the political boss Abe Ruef, an arrangement that might succeed in getting approvals to place unsightly overhead trolley wires throughout the city.

He thought, too, about his father and brothers. Never had a family been so conflicted in its business affairs or so entangled in family litigation. But Rudolph kept his concerns hidden, for he was fundamentally a silent, deeply thoughtful man, and safeguarded the internalizations of a lifetime, all his hurts and disappointments, harbored within a disciplined mind. He was the kind of self-assured gentleman whom Henry James would have regarded as an ideal subject—a lingering Victorian who found himself in a wild and wide-open society.

CHAPTER 4

Just eight blocks east of Rudolph's mansion was the home of Brigadier General Frederick Funston. The general was sleeping soundly next to his wife, a less imposing man in his nightshirt than in his medal-adorned uniform. Funston, at five foot four, was surprisingly short of stature to have achieved so mighty a position in the military establishment of 1906. His father, at one time a U.S. congressman from Kansas who had served with William McKinley, was much taller and more gregarious as well. Funston was quiet and reserved, a quite proper cut of the military cloth, and only on those rare occasions when he drank too much liquor did he ever curse or become aggressive. Funston was born in New Carlisle,

Ohio, on November 9, 1865, and the family's move to Iola, Kansas, a few years later, and then to Carlisle, Kansas, served to mold a typical midwestern farmboy personality. Growing up on the seemingly endless plains of America's farm belt mustered within him a yearning for the excitement of a bigger and fuller life.

Even his father's support did not help him when he tried, with very poor grades, to enter the U.S. Military Academy at West Point, and so he registered at the state college at Lawrence, Kansas. He remained there for two semesters studying botany, and though he left his formal studies without obtaining a degree he maintained a lifelong interest in plant life. A competent writer, he then took a job as a reporter for a strongly pro-Democratic newspaper in Fort Smith, Arkansas. In his later years Funston liked to tell the story of how as an ardent young Republican, he was asked to run the newspaper in the publisher's absence for a few days, during which time he wrote prodigiously in support of his own party. This reversal of the paper's editorial policy caused great consternation among the townsfolk and they, according to Funston, wanted to burn the newspaper down.

The young Funston returned to college in 1889 but again stayed just a short time, leaving in 1890—again without a degree—for a job in Washington, D.C., with the U.S. Department of Agriculture. He was sent on an expedition to the Dakota badlands to record flora and fauna. The challenges of the outdoors, the excessive and relentless heat and cold, the long hikes and rides over trailless terrain, the hunting for food, and all the confrontations that nature poses to the individual spirit greatly appealed to him. He went on a second USDA expedition in California to the scorching desert flats of Death Valley, where he camped out for eight months to discover and record many new species of plant life, insects, and animals.

He traveled to Alaska independently on another naturalist expedition, camping out, alone, on the bank of the Yukon during the winter of 1893–94, living off native plants and small animals, and enduring the ferocious cold through his own fortitude. After the snows melted that spring he built a boat, tested it, and then paddled more than a thousand miles down the Yukon, stopping to hunt and gather along the way, until he reached the Pacific. Once there he headed out to sea, aimlessly, knowing that if he stayed within a few miles of shore he would eventually encounter a passing ship. Finally, after several days, he found one, and it was headed for California.

From the boredom of his plains-state youth Funston had early developed an entrepreneurial and adventurous spirit, and late in 1894 he started a coffee plantation in Central America, though that effort failed apparently because it was insufficiently funded. Chagrined, he moved to New York and took a job as the deputy controller of the Santa Fe Rail Road, where he came to know both the railway system and railroad men.

In August 1896, Funston sailed from New York to fight in the Spanish-American War alongside the revolutionary soldiers in Cuba, where he comported himself with great distinction. He was shot in battle, captured by Cuban regulars, and tried and sentenced to death by execution. It was only through the intercession of his father's friend President McKinley that he was saved.

Fighting with the Cuban revolutionaries persuaded him to make his life in the military, for he loved its precise structures of human interaction—shoot or be shot in war and stand straight in a perfect row when on parade in peacetime. He was a thoroughly committed military man, and like most military men of the time he held a lifelong disdain for politics and politicians. However, the politics of one's personal life was a necessity that Funston was mindful of. When he returned from Cuba after eighteen months, he went on the lecture circuit, telling thrilled audiences of the rough and ready fighting that went on in Cuba and of his own gallantry. It was at one of his lectures, at Leavenworth, Kansas, that he was interrupted with a message from Governor Leedy, who was in the audience, asking to see him. Leedy had decided that he would muster four national guard regiments of Kansans to do their share in the war with Spain and wanted Funston to take command of one.

Funston and his men named their regiment the 20th Kansas Volunteers because there had been nineteen Kansas regiments that fought admirably in the Civil War, and they aimed to continue in that tradition. The governor then ordered the 20th to California to train for the war with the Philippine insurgents. Funston took his troops to the Lake Merced Military Reservation, just south of San Francisco, for five months of drill, but while they trained, the situation in the Philippines was changing rapidly. Emilio Aguinaldo, a Filipino nationalist, was forced to leave the country because of his constant agitation to Spain, but the United States supported him and equipped his troops while he was in exile, hoping that he and his freedom fighters would support an American invasion of the islands. When Admiral Dewey achieved his famous victory at Manila Bay he brought Aguinaldo with him, as the Americans believed he would make a good governor in the coming peace.

With the Philippines secured the men of the 20th Kansas prepared for an anticipated order to return to their homes. In the meantime, two things happened that were to change Funston's life forever: He met and married Eda Blankart and the government's foreign policy shifted. No longer was McKinley satisfied with a simple defeat of an enemy, for he became convinced by those in Congress who were warning of a "yellow peril" that it would be advantageous for the United States to have and maintain a strategic presence in the South Pacific. Accordingly, as was America's right according to the Treaty of Paris of December 10, 1898, McKinley bought the Philippines along with Cuba, Puerto Rico, and Guam from Spain for twenty million dollars. At about the same time Aguinaldo declared independence for the Philippine Islands, a move that took America by surprise, and the continuation of war was inevitable.

Funston had had good experience in Cuba in this kind of fighting, and unlike his fellow officers he was, because of his own reporting days, also adept at dealing with the press. This was not unimportant, for as the war grew more and more vicious, antiimperialist American journalists were quick to report all its gory details. The 20th Kansas was dispatched to the Philippines and distinguished itself by winning battle after battle, eventually becoming known as the Fighting 20th. They fought for eleven months in the tepid jungles and mountains of the islands, taking part in nineteen battles.

In one of them, Funston saw that he had to get his troops across the Rio Grande de Pampanga, a river that measured four hundred feet from shore to shore. Two of his men, William Trembley and Edward White, swam the distance under much gunfire, pulling towropes behind them. Funston then went in the first raft and pulled himself across on the towline, inspiring his men to follow. Soon, a sizable number of Kansans were in a position to attack the enemy's entrenched location and drove them into retreat. Funston, Trembley, and White were each awarded the Congressional Medal of Honor.

Still, there seemed to be no end in sight to the war, for the Filipinos were courageous and cunning, and their love of country propelled them with passionate intensity. Aguinaldo was a resourceful and formidable leader, and it was his capture that assured Frederick Funston of notice in the capital and the reward of the prized star of a brigadier. On March 21, 1901, when he learned where Aguinaldo was headquartered, Funston and a small band of his troops disguised themselves as captured American soldiers and were marched into Aguinaldo's camp by scouts posing as Filipino soldiers. Through this bravado, Funston was able to capture the rebel leader alive, which he did almost single-handedly.

Even after Aguinaldo was captured, however, America kept seventy-five thousand troops in the Philippines. There was much press censorship, and a propaganda battle ensued. Stories of American atrocities poured out of the war zone, filled with descriptions of concentration camps, water torture, and scorched-earth policies, and accompanied by heartrending photographs of trenches filled with victims. Mark Twain condemned the conflict as a "quagmire" while General Arthur MacArthur insisted that it was "the most legitimate and humane war ever conducted. . . ." In supporting the ultimate manifestation of the nation's manifest destiny, the Reverend Dr. Kirby of the YMCA told a convention audience, "The Lord Jesus Christ is behind the bayonets."

Although McKinley was assassinated on September 6, 1901, hostilities continued until March 1902. In the end, more than forty-two hundred American soldiers, twenty thousand Filipino troops, and two hundred thousand civilians died in the theater of what was called the Philippine Insurrection, the long and costly sequel to the Spanish-American War. The extraordinary number of civilian deaths could be attributed to the attitude of the American military toward the native population: They were in a war zone, they were in the way, they could not be trusted, and they were not of equal value as human beings.

This was the political environment that existed as General Frederick Funston returned a national hero to San Francisco earlier in that year. Many stories were written about his wartime record. He had proven himself to be a worthy fighting commander, and his capture of Aguinaldo had established him, along with President Teddy Roosevelt, as among the most famous and colorful of military officers of the time.

CHAPTER 5

Jack Murray of Engine Number 1 was always smiling, even as he climbed, once a day and sometimes twice, the merciless hillside trails on the Filbert Street side of Telegraph Hill. Telegraph Hill was not the highest hill in San Francisco, but its close view of the bay made it the likely place during the nineteenth century to keep a lookout for the revenue cutter service, which would fire off a cannon upon seeing the topsails of an approaching ship. This was announcement enough for the city's merchants, who would then run to the wharf to buy merchandise and dry goods from afar. The cannon was eventually replaced by a lookout who waved semaphore signals to indicate arriving ships, a sort of telegraphing that gave the 284-foot-high hill its current name.

Jack was thirty-seven, but he still took the hills as he had done as a child—in large, fast steps, each one taking him up the incline with determination. Sometimes, when he thought he needed the exercise, he would run up the 249 wooden steps on Filbert and Montgomery streets.

It would have been an easier trip home if there had been a cable car. But he knew a railroad would never reach to the top of Telegraph because the hill was so asymmetrical, with one side dropping off to a quarry and another a slope that was so steeply pitched it would have been impossible for even Andrew Hallidie to build a cable car to climb it. It was Hallidie, a Scotsman, who had built the first cable car in the world on Clay Street, in 1873. His invention had caught the imagination of the world, and soon cable cars were running in Europe and Asia as well as in other American cities.

Jack never lost his pace climbing the hill. His enthusiasm for everything came from the optimistic adventurousness that was his nature, from his conviction that you can never keep a good spirit down. He was handsome enough for any stage. A broad-faced man with deeply set blue eyes that studied everything, a prominent cleft in his chin, and a conservatively trimmed handlebar mustache that moved up and down as he laughed, Jack had that Irish wispiness that suggested he always knew more than he was saying.

Born in 1869 to Irish immigrants, Jack had grown up on the Barbary Coast, the sailors' leave district of the city, three blocks long and spotted with saloons. The family lived in a two-story frame house with a long front stairway, not far from the famous establishment that was fronted by a sign reading YE OLDE WHORE HOUSE or from other places where women called out from heavily draped windows unambiguous promises. If it was not a virtuous environment for a growing child, it was certainly San Franciscan in personality and a place where hard-working people got along well with the wild and lawless. It was here that Jack learned that the world's fortune was not going to come to his door; he would have to seek it out. And so he went to sea as a teenager, on the *Sea Wolf,* where he met Jack London and saw much of the Pacific. He sailed from port to port, taking on work wherever he could get it, until he decided to settle down and joined the San Francisco Fire Department.

The city's firefighters had just one day off a month and lived for the most part in their firehouses. They were given an hour for lunch and an hour for dinner, during which they could go to their homes to take their meals. But the Murray place, which stood near the very top of Telegraph Hill, was a good ten-minute walk, mostly uphill and at a fast and practiced pace, from the firehouse on Pacific Street. That left him just forty minutes with his family, but tonight Jack was taking both the lunch and dinner hour together, which meant an extra hour with Will.

At six, his son was at that age when a good amount of time had to be spent with him, when it is important that father and son come to understand each other clearly before any contrary ways set in. Jack had had a lot of time at sea to think about how he would raise a son and was convinced that they would have to get along well if ever the boy was to amount to anything.

He wanted it to be different for Will from what he had known as a child, when most of the working poor of the city lived in residences that would be called tenements on the East Coast, while the wealthier members of society built their homes on the higher and firmer ground of Nob Hill and Pacific Heights. Jack might have reached higher ground, but he was still far from achieving what he wanted—a place closer to the firehouse on Pacific Street so that he could spend more time at home. He was living high up on Telegraph Hill with many of the Irish working class and the new Italians, and he never gave a moment's thought to the luminaries who had been there before him, when Telegraph Hill was a small village for artists and offered its beautiful views to Bret Harte, Edwin Booth and his brother Junius, Mark Twain, Frank Norris, and even Ambrose Bierce—the same crowd that would later move downtown to the famous Monkey Block.

But it would not have been easy to find a new home with the wonderful space and expansive vistas he had at 340 Lombard Street. The inexpensive rent was also a factor, for just below him on the north and west sides of Telegraph Hill was a mountainside so cut up that it looked as if it had been bombed. It was there that Harry and George Gray ran the quarry of the Gray Brothers Artificial Stone and Paving Company, which sold gravel and stone for paving streets and garage floors, for seawalls and concrete mixture, and even for ballast. The Murray home might well have been the dustiest house in the entire city, built as it was almost on top of Gray's stone-crushing operation. Although he and his beloved Annie could see from one end of the great bay to the other—and on nights of the full moon, there was no better sight for any price in the world—the Murray family's living at the dusty top of Telegraph Hill was a relentless affront to Jack, realizing as he did that he couldn't afford a place on the flats, not on his weekly salary of twenty-three dollars.

Annie was waiting at the door as he arrived, and Will was just behind her, playing with his big fire engine. Josie, fourteen, was in the kitchen, stirring a stew, as Anita, twelve, was setting the table with nine-year-old Myrtle. Estelle, just crawling at one, sat on the floor with alphabet blocks.

Annie threw her arms around her husband, as she did each time he came home, each time wishing it was his day off and he could stay the night. Born Anne Jordan in San Francisco, she had full Irish lips, and was slight at five foot three, but the way she pulled her hair back in a mesh of brunette curls always made her look taller and straighter. Jack held her hand as they walked to the kitchen, a hand small and firm, so unlike the softness of her thin hips.

Will got up and ran to his father. "Paw," he said, "I want to take the fire truck to school, okay? Okay?"

The fire truck was more than four feet in length and had been made especially for Will by Mr. Graves, who owned the carriage-building shop next door to the firehouse. It was a special gift that had come the previous Christmas, and Jack never imagined a boy could love anything as much as Will loved that big fire engine.

"You leave your father be," Annie said, " 'til after his dinner."

"It's a big fire engine, Will," his father answered, "and we would have to hire a truck to get it there."

"No, Paw. I can wheel it."

"Leave your daddy alone now," Annie repeated. While she found it insufferable that her son alone among the children called his father "paw," she realized it was simply another sign of the child's independence.

Jack chuckled to himself as he threw his fireman's vest over the back of his chair and tickled each of the girls. With five children, there was never a peaceful moment, let alone the fact that department rules required that each firefighter place at his own expense an alarm bell in his home. As he did each time he sat for a meal, he looked at the six-inch brass bell on his kitchen wall and hoped it would not ring while he was eating. Jack had originally wanted an eight-inch bell, but Annie protested; her brother was a fireman in Oakland and she knew how loud the larger size could be. "This is my kitchen," she announced, "and there will be no bell bigger than six inches across."

After dinner, Jack sat in the small living room and found an answer to the dilemma of his son's fire engine. On his next day off, Jack explained, he would be able to help Will carry the fire engine down the hill to his school. He couldn't help but admire the way Will was determined to show all his classmates his proudest possession.

As Jack began to make his way back to the firehouse he passed the Italian neighbors who lived just across the street and exchanged pleasantries. These were people who kept their noses to the grindstone, Jack thought, and though nearly the same in age as him and Annie they already owned their house, something Jack only dreamed about. They might have gotten a loan from that new bank, the Bank of Italy, which some of the Italian firemen from North Beach had been discussing. A. P. Giannini, its founder, was said to lend money to working people to buy houses and open businesses, which made his the first bank in the city that did not deal exclusively with the millionaire class—as did Wells-Fargo, the American National Bank, and the First National Bank of San Francisco.

The big wooden doors of the Engine 1 firehouse at 419 Pacific Street were spread open as he reported back to duty. It was a plain red structure, very unlike the more ornamented firehouses in other parts of town that Jack remembered from his youth. These were the holdovers from the volunteer days, like the St. Francis Hook and Ladder Company 1 on Dupont Street, with its Grecian pillars and pallisters; the Pacific Engine Co. 8 on Front Street with its life-size sculptures of proud firemen standing on its cornice; and the Monumental Engine House on Bernham Place with its high bell tower.

Captain Murphy—known as Big Tom to distinguish him from the other Thomas Murphy in the department—was standing in front, his large, pugnacious hands placed on his hips just below the woolen fireman's vest. The captain gave Jack a generous nod as he appeared, indicating his approval of Jack's punctuality. The captain was not a gregarious type, but quiet and studious—some would say hard-nosed. In just four more years, because of his connections with a radical change in government, he would jump over all of the city's other top chiefs to become the chief of department. It would be an honor he would hold for nineteen years.

Just then Jack saw Mr. Graves and his daughter walking out of his shop. As they crossed the high and heavy cobblestones fronting the firehouse, Jack called out, "Mr. Graves, that fire truck you built for Will has taken over our household!"

The Graveses laughed and stopped for an idle talk. Jack then introduced the comely Annie Graves, who worked as a stenographer for her father, to the captain, little knowing that the introduction would lead in a year's time to her marriage to San Francisco's future fire chief.

A few of the boys—Kelly, Capelli, and Tyson, all of whom had taken a long lunch—were in the back room of the firehouse, a kitchen of sorts, having a sauerbraten dinner prepared by the German hoseman Auggie. Jack spoke a few words to the Berliner and then checked the coal and water levels in the steamer, for he was the temporary driver that night. He then retired to his bed in the bunking room on the second floor, where a small gaslight there provided just enough illumination to read by and went through the day's newspapers.

One by one the firemen climbed the stairs as the evening wore on and joined Jack in the sleeping quarters. They removed their uniforms and, in the wavering glare of the gaslight, folded their pants and stuffed their socks into their boots at the side of their beds, preparing for the next alarm.

There was not much local news, Jack thought, flipping the pages. The headlines emblazoned a brutal lynching of three Negroes in Springfield, Missouri. There was a further report on the volcanic eruption at Vesuvius the week before, which had devastated Naples, Caruso's home town. Another story detailed the great generosity of San Franciscans in raising relief funds to send to the victims of the eruption. The newspapers gave so much space to Vesuvius, Jack suspected, because it had once buried the town of Pompeii in nine feet of hot ash, killing more than two thousand. Jack had read Pliny's account of this tragedy while at sea and remembered it well. He also recalled that just one year before, on April 4, 1905, an earthquake had struck Lahore Province in India and killed more than nineteen thousand people, but the papers did not give that event much space at all. Nor had there been much coverage of the forty thousand who were killed when Mount Pelee exploded in Martinique on May 8, 1902.

As he went on reading he saw that a costume ball was to be held that night at the roller rink, at the Mechanics Pavilion on Larkin and Van Ness streets and the prize for the best costume was a diamond ring for the lady and a gold watch for the gentleman. Wouldn't Annie like a diamond ring? Jack thought. His interest in the sea had never waned, and he turned to the shipping column. He saw that the SS *Sonoma* was leaving for Sydney in two days and the *Acapulco* was going to New York in three. A lost and found notice described a lady's watch with three fleurs-de-lis in tiny diamonds gone missing in the vicinity of the opera house, and promised a reward of twenty-five dollars for its return. Jack laughed, for what was a keepsake to one person was more than a week's salary to another. Finally, he closed his eyes and thought of Will's presenting his fire truck to his class at the Union Grammar School. It was his final smile of the day, and he was soon asleep.

CHAPTER 6

Many San Franciscans believe that the Bay Area, if not the bay itself, was discovered by Sir Francis Drake in the late sixteenth century. For years historians have searched for the famous "plate of brasse" that Sir Francis said he nailed to a tree in 1579 at 44 degrees latitude on the west coast of North America, thereby claiming the land, which he called New Albion, for his queen and country. Today, a small etched brass plaque hangs in the lobby of the Bancroft Library at the University of California at Berkeley. It was discovered in 1936 in a Marin County woods and was believed to be the archaeological find of the century. The problem is that, in 1977, when the plate was subjected to a neutron activation analysis it was found to contain very high concentrations of zinc, a metal that had not yet been identified as useful in the sixteenth century, and likewise found to have been modernly pressed and not antiquely hammered. The forgery simply fueled a new controversy regarding Drake's visit to the shore of California.

Modern historians have hypothesized that Drake mistakenly believed that he had found near what is now called Vancouver Island a northwest passage from the Pacific to the Atlantic. Because the queen, Elizabeth I, wanted to keep this supposed discovery a secret, she had Drake falsely identify his route in his ship's log, claiming that he was much farther south, all the way to the coast of what is today Marin County at 38 degrees latitude. This would prevent the Spanish from getting an inkling of his new discovery, and would give Great Britain a trade route that would allow it to circumnavigate the powerful Spanish fleet. It has also been argued, moreover, that Drake's New Albion was really an inlet on the Oregon coast called Whale Cove, which is at 44 degrees latitude, a position that has been corroborated by a document now in the British Museum written by a man said to be one of Drake's crew.

So if it cannot be firmly established that it was Sir Francis Drake who discovered San Francisco, that distinction must be granted to a lowly sergeant who in 1769, on an overland expedition led by Gaspar de Portolo, looked up from a mountain peak and saw the great bay in the distance and dutifully noted his observations.

But the glory of actually claiming San Francisco was to fall to Captain Juan de Ayala, who was the first European to sail into San Francisco Bay on August 4, 1775. He made an encampment there for forty days, surveying the area for the Spanish crown and for the priests like Father Junipero Serra, a Majorcan member of the Order of St. Francis, who was creating missions up and down the coast of California. These missions would later become cities, like the one he named El Pueblo de Nuestra Señora la Reina de los Angeles de Rio de Porciuncula, or the Town of Our Lady the Queen of the Angels by the River of a Little Portion—today's Los Angeles.

A famous diarist of the time, Padre Font, a Spanish cleric on yet another Spanish expedition later in 1775, found the bay, sixty miles long at its greatest extent and fourteen miles wide, to be "a marvel of nature . . . a harbor of harbors." In October 1776, not far from its shore, a priest named Francisco Palou founded San Francisco de Asis, the sixth mission to be built in the series of twenty-one missions that made up El Camino Real.

It can be said that the real impetus for the growth of San Francisco was Ferdinand Magellan, who in 1511 made a proposition to Portuguese speculators that he could circumnavigate the world if they would buy and equip fifteen ships for the adventure. Magellan ultimately discovered the only passage through Tierra del Fuego on the southernmost tip of South America, where, with an extraordinary and extremely dangerous rocky outcropping on each side, the Atlantic and Pacific oceans meet. More than three hundred years later, would-be miners and purveyors of services of all kinds would use this route to transport themselves and their equipment around South America to jump-start a gold rush that not only populated a new city, but also defined its image as rough, wicked, enterprising, and fabulously successful.

But there had to be a better way to get from one coast to the other than a monthslong, wearying sea voyage. In 1845, John O'Sullivan, writing in *United States* magazine, established a national goal and in consequence coined a term remembered by every student in their first American history course: ". . . [it is] the fulfillment of our *manifest destiny* [emphasis added] to overspread the continent allotted by Providence for the free development of our yearly multiplying millions."

Of course, manifest destiny dovetailed nicely with the rush for gold, but after the Civil War there was in general a renewed interest in America in moving westward, which would be sustained until 1906. After the Irish and German immigrations, the New York tenements continued to bulge with new arriving groups like the Italians and Eastern Jews, and many of them sought to escape the overcrowding and meager work opportunities of the East Coast. If America represented opportunity to immigrants, then the American West represented opportunity plus space. New horizons began to open as the Santa Fe Trail was linked to the Old Spanish Trail, to the Oregon Trail, to the Mormon Trail, and then to the Oxbow Route from Missouri to California. Brave and adventurous Americans pushed to the Appalachian Mountains, then to the Great Lakes, the Mississippi Valley, the Rocky Mountains, the Sierra Nevada, and finally to the Pacific: San Francisco, Los Angeles, and San Diego.

In 1849, the population of San Francisco was about 5,000; by the beginning of the Civil War it had risen to 57,000, among whom were 3,000 Chinese and 2,000 African Americans. After the war the population doubled to more than 100,000 and was the tenth largest city in the nation. By the turn of the twentieth century, 450,000 people were living in San Francisco.

But the overland treks brought their own great hardships, and an easier way west was sought. The transcontinental railroad was completed in 1869 when Leland Stanford drove with a silver hammer the last, and a golden spike at Promontory, Utah. But rail travel remained expensive, particularly for moving freight. The government had discussed for decades the possibility of cutting the isthmus of Panama in half with a canal, and, after supporting a revolution in Colombia to pave the way to the land's acquisition, President Roosevelt finally began construction of this engineering wonder in 1904. Yellow fever and malaria did much to slow the work, but it was finally completed in 1912, and dedicated by President Woodrow Wilson.

American expansion to the West culminated in the administration of Theodore Roosevelt, whose own personal qualities of hard work, independence, and rugged individualism suited the character of the booming western cities, while his love of the open landscapes helped preserve much of our natural beauty within his new National Parks System. Adventure, discovery, opportunity, and possibility were the stuff of the West at the turn of the twentieth century, and particularly of San Francisco of 1906.

CHAPTER 7

The foundation for the great wealth of San Francisco was laid when an army of adventurers, greedy con men, failed farmers, eastern Brahmins, European immigrants, neophyte prospectors, and experienced miners rushed to California in 1849 to seek wealth in the newly discovered gold fields. The population of San Francisco rose from five thousand that year to more than one hundred thousand by 1865. Though no single individual made a personal and lasting fortune from gold (even Thomas Sutter of the famous Sutter's Mill died a broken man), the great influx provided the basis for a growing merchant class and for the integration of San Francisco into the burgeoning farming and manufacturing export business to Asia, making the Bay City the premier Western port in the United States.

The famous fortunes of San Francisco—those that have become known as the big four—had their origins in the engineering plans of Theodore Judah for a railroad through the American West, the most rugged territory in the country. In 1860, four Sacramento businessmen—Amasa Leland Stanford, Collis B. Huntington, Charles Crocker, and Mark Hopkins—pooled a modest amount of money, just $7,000 among them, to create the Central Pacific Railroad. Because the federal government had an interest in attaching California during the Civil War, both to secure the growth of the country and as sort of a geographical investment should the South prevail in its determination to leave the union, the Central Pacific successfully pursued a governmental subsidy to connect the two coasts by rail. To accomplish that, they capitalized their new company with a fictitious $8.5 million. The government, indifferent to the due diligence standards required by any business, never checked their financial position, and so the big four were allowed to solidify their company and sell its stock. Then with brazen self-interest, they created the contracting company to do the work, and at the same time excluded Judah, who died of yellow fever while crossing the isthmus to make his very justifiable complaint heard in Washington. The big four also fraudulently surveyed an additional twenty-five miles of fictitious mountain range that required ties and rails, and in this way increased the value of their subsidy contract by $400,000. Such chicanery had its political rewards, and in 1861, Leland Stanford became governor of the State of California.

Construction of the railroad began in January 1863 and was completed that historic day in May 1869. Within a very short time these four men became among the wealthiest in the world. Stanford went on to create one of America's great universities, Huntington's money funded the formidable Huntington Library and Art Collection, and Crocker went on to create the Crocker–Woolworth Bank, which evolved through his son William into the Crocker National Bank. Mark Hopkins (who had nothing to do with San Francisco's famous Mark Hopkins Hotel) made a great deal of additional money by bartering iron and steel products throughout the Western states.

While the families of the big four would rise to become among the most socially prominent figures in American society, the same was not true for the Irish big four of San Francisco, otherwise known as the silver kings. These men, all born in Ireland, made their money in an atmosphere of shovels and picks and dirty Levi Strauss overalls, which made them less socially acceptable than the railroad men, whose heads were kept high by stiff linen collars tied with silk.

James C. Flood, William S. O'Brien, James Gordon Fair, and John W. Mackay might have been proletarian millionaires, but in the long run they could have bought and sold the railroad swells. Their legacy had its origin in the fact that Flood and O'Brien, who were partners in a bar and grill establishment called The Auction, were good listeners and put together bits and pieces of information they gathered from the imbibing miners down from the Nevada hills. In 1859, they asked Fair and Mackay, who already had some mining success, to join with them in setting up a stake in Virginia City, where the prospects were thought to be certain, or at least worth betting on. They bought into several mines and by clever bartering found themselves in control of the Consolidated Virginia Mine, which harbored the Comstock Lode, the largest deposit of silver ever discovered before or since—a fifty-foot vein that spread far enough through a mountain to yield more than five hundred million dollars in pure ore until its depletion in 1869. (In today's value, that amount of money would be on a par with the great fortunes of Bill Gates and the Walton family.)

It can be fairly asked if the silver kings had little social success in San Francisco because they were immigrants, men with the singsong brogues, the first generation off the boat. A proper place in the social order required a little more refinement than they could summon, but these men had filled their lives with work and had little time to develop sophistication with knowledge of literature and art.

In fact, San Franciscans were not prejudiced against the Irish, as was the case in New York and Boston. The city prided itself, then as now, on its liberal acceptance of all people. The Irish generally fared well in the Bay City, as did Jews. Such families as the de Youngs (changed from DeJong), who owned the *Chronicle,* the Hellmans of Wells Fargo, and the merchant families of Levi Strauss, Dinklespiel, Schwabacher, Fleischhacker, and I. Magnin were all socially prominent.

If there was significant prejudice in San Francisco it came in the form of nativist concern about the continuing influx of Asians and the ever-growing pool of cheap labor it provided to compete with more settled workers. One of the most prominent of these nativists was Rudolph Spreckels's partner, the three-term mayor of San Francisco (1897–1902), James Duval Phelan. He was the son of James Phelan, an Irishman from County Cork, Ireland, who arrived in New York as a boy. It was said of the elder Phelan that as he grew into manhood everything he attached himself to made money. When he read of the discovery of gold in 1848, he immediately recognized an opportunity, not to make a mining fortune, but to supply the miners. He filled three ships with merchandise of all kinds and sent them south around the Horn, bound for San Francisco Bay. One of the fleet sank, but two ships arrived with enough cargo to enable Phelan to establish himself as an honest merchant. In twenty years he prospered so well that in 1870 he could finance the start-up of the First National Gold Bank of San Francisco and become its first president.

His son was to succeed him as president and was just the kind of Irishman who naturally endeared himself to everyone he met. Had it not been for his vehement and nativist anti-Japanese sentiments, he might have gone on to be the most beloved politician in the state's history and despite them is still known as San Francisco's first honest mayor.

James D. Phelan was definitely a rich man's son, but he was determined to make his own mark and was so well placed in society that he found many opportunities, eventually becoming president of the Mutual Savings Bank, and then chairman of the United Bank and Trust Company. His good friend Rudolph Spreckels was president of both the First National and the United Bank and Trust Company and also served on the board of the Mutual Savings Bank.

Phelan was a natural leader and had a complicated intelligence, a sort of man for all seasons who was able to write courtly love letters in the morning, resolute business instructions in the afternoon, and political apologia in the evening. He was also, notwithstanding his nativist distrust of the Japanese, a highly ethical and committed public servant.

Three things had been of great concern in this civic-minded gentleman's life in recent times: the City Beautiful plans, the Hetch Hetchy water proposal, and the fight against corruption.

Before he ran for mayor, Phelan visited the 1893 Colombian Exhibition in Chicago and was inspired there to a life of civic improvement. It was also there that Phelan toured the White City, a collaboration of prominent architects and artists of the time, led by the architect Daniel H. Burnham, and saw the possibilities of sound municipal planning. This idealized city built entirely in white gave rise to the City Beautiful movement and was the major attraction at the fair. Phelan knew that a city that didn't build would soon face its own demise, and he could see the possibility of the structures that might be built in San Francisco, none of them more than sixty feet high, bordering expansive green belts that would inspire a generation.

During his mayoral administration, Phelan hired Burnham to design a City Beautiful plan for San Francisco that would replace the hodgepodge of construction that had arisen without consideration of an overall plan. Most buildings were jigsawed and scalloped structures cheaply built of wood to, it was argued, sustain earthquakes. As the nineteenth century passed and taste turned toward the antique these buildings would become known as the prized painted ladies of San Francisco. But in entering the twentieth century Phelan envisioned monumental palaces of masonry and marble set in Arcadian simplicity, low and elegant and classical, buildings that would sustain earthquakes as the Grecian temples had for more than two thousand years. He saw fields of flower-bordered grass, running in long, wide avenues from one end of the city to the other.

Burnham decided to design the City of San Francisco after Georges Haussmann's plan of Paris, with concentric rings extending out from the waterfront. The plan was presented, but the money to build or restructure parts of the city was never found.

But of all the imminent improvements needed in the city, Phelan understood that water was a more fundamental problem. The Romans had inscriptions placed on their viaducts reminding the citizens of the might and genius of their leaders, resourceful and thoughtful men who were able to bring them so abundant a water supply, the most abundant in the ancient world. In San Francisco, however, the private ownership of a limited water supply, Phelan knew, could only undermine the growth potential of the gem city of the Pacific. Instead of paying homage to its civic leadership for its water, San Franciscans of 1906 worried about how the city would be able to supply water to an expanding population.

Surrounded on three sides by an endless supply of salt water, an inadequate supply of freshwater was a constant problem for San Francisco from the beginning of its existence. It could not count on a predictable amount of rain, for it was subject either to drought or to unremitting floodlike downpours, and the annual rainfall of fewer than twenty inches was a constant reminder that the city needed a new water supply to build its future. Its sandy soil could not be counted on to carry streams or creeks. Few city parks had adequate water, so the ground coverings were more brown than green, and only a very few wealthy home owners could afford water enough to create and maintain the green in their lawns.

In 1859, an entrepreneur named George Ensign had obtained from the California legislature the right of eminent domain to confiscate land for the development of a water supply for the common good. It was a license to build a powerful land company as well as a water supply system. The lawmakers had already delimited the size (and the potential of growth) of San Francisco by lopping off 80 percent of what was then San Francisco County and creating San Mateo County, leaving the City of San Francisco and the County of San Francisco the same size, just as they are today—there is not an inch of land for potential expansion in the county.

Ensign hired a German engineer named Alexis von Schmidt to bring water from San Mateo to San Francisco, thereby acquiring a monopoly that was to hold for almost fifty years. But von Schmidt later abandoned what became the Spring Valley Water Company to build his own water system, the Bay Cities Water Company, which would draw from the Lake Tahoe area, 160 miles away, and promised to deliver four times the five million gallons that were being provided to San Francisco by Spring Valley. But von Schmidt had not counted on the newly created State of Nevada objecting to his drying up the Truckee River and despoiling the beautiful Sierra Nevada landscape. His scheme went bankrupt, but it was about to be resurrected in opposition to James Phelan's civic-minded plans.

In 1900 when Phelan was mayor, he introduced and fought for a charter reform that freed the city from the control of the state legislature and empowered San Francisco to meet its water problem directly. The reform also established a board of public works that, at least under Phelan, would act in the best interest of the city, taking the place of a state legislature that heretofore had accommodated Spring Valley's water monopoly. Phelan studied all the possible water sources from the most reliable engineers. The stranglehold that the Spring Valley Water Company had on the city had to be disengaged, and the city would have to own its own water to compete both in commerce and as a socially dynamic place to live. Phelan concluded that the Tuolumne River and the Hetch Hetchy Valley to the north presented the best source for the city's water supply and reserve, and in 1901 he began as a private citizen buying up water rights and land in that area, knowing that these private acquisitions could be kept secret, thereby circumnavigating the rape of the land profiteers. He had also made an agreement that he would sell the city the rights to the land at cost plus nominal interest.

There was a significant obstacle to Phelan's plan, however: The valley that had to be dammed and filled to create a great new reservoir was part of the Yosemite National Park system. It was a disappointment to Phelan when, in 1903, the secretary of the interior declined to approve his 1901 application to create a new water supply, arguing that the secretary had no right to give up national park land for eminent domain purposes.

This left the question of water up in the air, so to speak, and the new mayor, Eugene Schmitz, received many ideas and applications for new water companies to compete with Spring Valley. But now, as many people were beginning to realize, the six years of the honest reform government of James D. Phelan were over, and the whispering among the city's businessmen had already begun. Mayor Schmitz and the burgeoning Union Labor Party were for sale, city boss Abraham Ruef was their agent, and a new water system became a potential prize in their crooked portfolio.

In 1906, twelve hundred landowners made application to Mayor Schmitz to abandon the Hetch Hetchy project that Phelan had championed. These applications were from obviously self-interested citizens and were to be expected, but they were joined by protests from the Sierra Club, a formidable opposition. At virtually the same time the Phelan plan was rejected, the mayor received eleven formal applications for proposed water systems, among them a proposal from the Bay Cities Water Company to bring water from Lake Tahoe—its rights newly acquired by William S. Tevis of the Tevis Land Company. Tevis's success as a real estate magnate also enabled him to offer a one-million-dollar bribe to Abe Ruef to persuade the mayor and board of supervisors to buy the water rights to the Bay Cities Water Company for ten million dollars. These were the same rights Tevis had offered to the city the year before for two hundred thousand dollars but the stakes had been raised with the abandoning of the Hetch Hetchy plan. Ruef had planned to spread a half million dollars among the eighteen supervisors, and he and Schmitz would split the other half.

CHAPTER 8

Bridget Conroy loved to spend the night with her little ones, her grandchildren, Coleman Junior and Helen. Her son Coleman and his wife would be away until the morning, and they had asked Bridget to sleep over that Tuesday night in their home at 855 Harrison Street. Coleman was in the liquor business with a distributing company on 3rd and Verona, and, by Bridget's assessment, was every bit a successful man. She regarded these evenings she spent babysitting for Coleman's children as a special gift, for they reminded her so vividly of the accomplishments of her long life, from her childhood on a meager farm in Galway to enduring those terrible, suffering years of the potato famine to sleeping overnight in her own son's beautiful home in San Francisco, a house that only an English lord could afford in Ireland.

The Conroy family, like many immigrants of the time, had responded in the early 1880s to an advertisement in an Irish newspaper: Come live in Minnesota, and Minnesota will pay half your travel. Destitution was still widespread in Ireland, particularly in western Ireland's Rosmuc, where they had a farm, and so the offer of a shared expense to America was an incentive, perhaps even an opportunity. It would be difficult, of course, because like most Irish immigrants of the time, they were Irish speakers, but they knew some English and would get by. And so Bridget's family left for Minnesota, where they found that it wasn't language that made them feel insecure but the unrelenting midwestern cold, which was more than any in the family could endure. The family decided to return to Ireland, all except Bridget and her brother who decided to head for more temperate climes in America. Now, twenty-five years later, she was a regular San Franciscan.

Easter was coming up, and she had told the youngsters the story of the wren song, which all children in Ireland sang to their neighbors on Easter Eve with wry hopefulness for a penny in reward. It had been a long day. Weary with fatigue, perhaps with age as she neared seventy, she kissed her grandchildren and put them into their beds.

Though just a few streets north from her own two rooms at 118b Juniper Street, her son's house was in a world very different from her own, which was, really, more an alley than a street. Bridget would not complain. No, never. Her marriage might not have been made in heaven, but there were the eight children, healthy and enterprising. Her son Patrick stayed on the farm in Rosmuc and had thirteen children, and most of them had come to San Francisco. They were all dutiful and hard-working, enough to make good lives for themselves. It was a typical immigrant story, the saga of the Conroys. It had been better than finding gold on the streets, for there was nothing better than to have a loving and successful family.

In the small room next door to the children's bedroom, Bridget knelt on the thin Persian rug. She was so tired that she had no thought in her mind other than to get through her night prayers. Praying was what Irish girls did, she thought as she shifted into the thin cot, the lucky ones and the unlucky ones as well.

CHAPTER 9

Henry Lai squirmed in his seat. Sleep, and even rest, was next to impossible on the hard, thin-cushioned seat. It had been a most uncomfortable trip on the connecting railroads from Cleveland. But it was a small sacrifice, he reminded himself, to travel on the most inexpensive ticket possible, for he had his letter of recommendation from the Cleveland Presbyterian ministry, and he was soon to meet his beloved Yuen Kum in just hours. His train was due to arrive in Oakland at 6:00 A.M., and he would travel from there to the Chinatown Presbyterian Mission at 920 Sacramento Street where Yuen Kum was in the care of Miss Cameron.

Henry, a slight young man with highly accented and limited English, thought he could never be happier than that day a few weeks earlier when he had received the letter from Miss Cameron saying that she approved of his letter of proposal to Yuen Kum. But now, his meeting with his love just hours away and his wedding ceremony only a few days beyond that, his happiness was even greater. It had been more than six months since he had first interviewed with Miss Cameron and met Yuen Kum, hoping he might be able to arrange a marriage. All of his dreams were coming true, his time in America was truly becoming a *gold mountain,* which is how the Chinese thought of America and referred to it in their poetry.

At that moment in San Francisco's Chinatown, Donaldina McKenzie Cameron was as always ministering to her girls, caring for their persons or their spiritual needs, putting the young ones to bed, leading them in their ablutions and their prayers. She had fifty Chinese and Japanese charges in the Sacramento Street mission house, and all of them had worked so diligently sweeping, mopping, dusting, and shining that day to make the mission house ready for the following morning's annual meeting of the Occidental Board of Foreign Missions, and for Yuen Kum's wedding, to be held on Sunday. They were proud of their mission house, and most of them knew how important the annual meetings of the board were for Donaldina. They wanted things to be in perfect order for her. They were excited for Yuen Kum who was actually living the one delightful fantasy they all shared.

Finally, they were all in their bunks—the *mooie-jay,* the little girls who had been used as household slaves, and the older ones, who had been rescued from a worse kind of slavery. Yuen Kum, being the oldest, was in the first bunk of the final room of Donaldina's nightly rounds. In one high-voiced chorus the girls sang out, "Good night, Lo-Mo."

Donaldina smiled, as she had done each night for more than eleven years with hundreds of different girls, perhaps thousands. "Good night, Mother" was an address of respect among these unfortunate children, sold by or stolen from their families. They had been shipped to the United States to find themselves in the most degrading situations, and after constantly beseeching their ancestors for a miracle had their prayers answered by their Lo-Mo. Donaldina did not return their wishes for a good night, because it implied an ending of a kind, even if only the ending of a day. She was providing a refuge from a terrible alternative for her charges, and it was vitally important that there was no end of any kind to that safe harbor. She appreciated the comfort of that safety, for her own family had come to America in 1871, when she was only two, after her father had lost a fortune in a Scottish financial crash.

"Sleep well," she said.

It had been a long day, and Donaldina finally fell into a large, soft chair in her room and reached for her Bible. She loved falling asleep with the holy book in her hands, turning the pages until she came to her favorite passage: *I will never leave thee, nor forsake thee.*

CHAPTER 10

Jack could count the bells even in his sleep. First one, then nine, followed by seven. Box 197. He knew that box, at the corner of Bay and Mason streets, just a few blocks from the docks. Lot of warehouses over there, he thought. He looked at the wall clock. Eleven o'clock. He had slept for an hour, maybe.

In seconds, Jack was in his clothes and sliding down the thick brass pole. Since he was assigned the position of driver he had a lot to do. He could hear the horses, which knew the boxes as well as he did, and though they usually slapped their shod hoofs against the stone of their stalls, always ready to run, they now seemed to be screaming. Someone else would have to calm them and lead them up, because Jack had to fire the 1893 Amoskeag steamer and had little time to do so if they were to get out of the firehouse in ninety seconds. He gathered up excelsior and sulfur with which to make a compound of Greek fire, pushing the mixture beneath the coals, and stepping back a little as it flared when a match was laid to it. Joe O'Brien, the stoker, came with a shovel in his hand and began to load the steamer's fire compartment with coal, and they both could hear Tim Collins singing the word "easy" over and over as he began to guide the horses, the two bays and the white one, to the front of the steamer. Jack pulled down the automatic collars and the horses, which normally received them with equanimity, now balked, and he had to push and pull them into the harness. Even so, they were eager to run, as charged by the excitement of an alarm as any fireman or firehouse dog.

It would not be long before the steam had built up enough pressure to push water at five hundred gallons a minute. Jack stopped to check the two eight-foot-long, four-and-a-half-inch diameter suction hoses that lay one upon the other at the side of the steamer, making sure they were secure. A pumper is of no use if there isn't a suction to connect it to a reservoir of water or to a hydrant. He then checked the quick connection to the hot water hose that ran from the firehouse boiler to the steam engine itself to make certain it was working, that the steamer was filled with hot water, and that the connecting hose would fly away as designed.

Jack climbed into the small leather seat at the front of the steamer and positioned himself firmly in its cupped profile. He was high up on the vehicle, his feet now level with the backs of the horses. He looked around the firehouse and saw Captain Murphy stepping up onto the hose wagon, just as he saw O'Brien on the back of the steamer waving the go-ahead. Jack made certain he held the reins firmly in his left hand. With his right, he pulled a long black whip back and then cracked it over the horses' heads. The hot water hose flew off its connection as the steamer lurched forward, and Jack held steady as he began to feel the wind. The whole turnout took no more than twelve seconds—they occasionally did it in nine. The three horses were at a full run as they reached the narrow door of the firehouse and made the close turn up Pacific Street to speed around Telegraph Hill. O'Brien was pulling madly on the bell rope to give sufficient warning to any pedestrians who might be wandering through the streets.

At least it's a flat run, Jack thought as they sped to the fire, and we won't have to push and pull the steamer up any hills.

They saw the clouds of smoke and the fierce glow pushing upward in the warm sky as they turned down Old Montgomery Street. It was the lividness of the orange glow that prepared them for the action, because nothing but a heavy burn fills the sky with a color so intense. Jack felt the wind fully now and the blood rushing through his veins. It was a job, and a big job at that.

Chief John Dougherty, the department's assistant chief, arrived just as Jack pulled the reins back to a stop at the corner of Bay and Taylor streets and began to search for a hydrant. Spotting one he moved the horses closer to the fire and began to connect the suction from the hydrant to the steamer. He heard Chief Dougherty yell to his operator, "Go to the box and tap in a third alarm on arrival." The department's telegraph, its only communications system, had never failed, and Jack knew that there would soon be six additional steamers on the scene as well as a few more truck companies. Even the boss, Dennis Sullivan himself. This would be a major effort, for the California Cannery Company's fruit warehouse, a building half a block long, was completely afire. The heat was great, and the smoke was swirling up and down before them, but the firemen were used to this radiation and crouched lower as they aimed their hose.

In a short time, more than 120 firemen—more than 20 percent of the city's entire force of 568—were fighting the cannery fire from every angle. Little by little they made their advance, keeping their water streams in circular patterns to break up the heat, until they were at the axed doors and broken windows, and then in the black interior.

Jack suddenly had an ominous thought. It had been in a building just like this one only ten weeks before, on the night of February 1, 1906, that Charlie Dakin, the captain of Engine 4, and T. J. Hennessey, the hoseman of Engine 22, had been trapped in a collapse and had perished in the choking smoke. There had not been a mark on either of them, and every firefighter understood their friends had choked slowly and in painful recognition of an impending death. That memory was so fresh that Jack wondered as they began to enter the dark of the building if Dougherty would let them continue an interior attack.

As soon as he arrived the fire chief, Dennis Sullivan, surveyed the blaze with concern. Chief Dougherty had immediately called out a third alarm, and the fire was in one of the most crowded warehouse sections of the city. The buildings stood so close to one another that the interior heat alone would be sufficient to fire up an adjoining structure. In firefighting, an imminent catastrophe is always possible, but at least they were on the flat and close to the water. He would not have to worry about a dry hydrant in this location.

Jack Murray watched Brownie lead Chief Sullivan's buggy to the front of the fire building, prancing and with head held high. Brownie was known in the department for getting closer to the flames than any other animal, horse or dog, and Jack loved to watch the proud and muscular horse every time Chief Sullivan appeared at a fire. An Irishman, he thought, seemed to have the love of horses in his blood, and the department's horses, particularly—half Percheron dray horse and half racing trotter—were more than worthy of admiration. They were specially bred for the fire department by Ben Ali Haggin and Lloyd Tevis at their horse farm in Bakersfield. Indeed, it was said jokingly that many Irish immigrants joined the fire department because they came from farms and missed the back end of a horse. These animals cost about three hundred dollars each, and about 325 of them were spread throughout the fifty-eight fire companies in the department's forty-eight firehouses.

Jack left the operation of the steam engine to Joe O'Brien and joined his company at the entrance to the building. Captain William Nicholson and the men from Chemical 3 were there as well giving a hand. The substance in their tanks, carbon tetrachloride, would not be as effective as water in a big-volume fire like this, so they were pulling hose wherever it was needed.

Jack watched the action before him as he relieved the back pressure on the nozzleman by holding high the two-and-a-half-inch cotton hose. He also kept an eye on Dennis Sullivan because Sullivan, "a chief among men by nature," was a legend in the job, and to watch him in action was a little like watching the president of the United States or the king of England. To Jack, there was no greater honor in all the world than to be the chief of the San Francisco Fire Department.

Chief Sullivan immediately took command and consulted with Chief Dougherty. Soon, Chief P. H. Shaughnessy joined them, and the three, the iron triangle of political and intellectual power of the department, walked quickly to the side of the building, toward Francisco Street. They were checking the exposures and the walls, Jack realized, and he wasn't surprised when Chief Sullivan ordered all the firemen out of the building and off the roof. Removing them meant that it might take a little longer to get the blaze under control, but Jack saw the wisdom in his decision. Chief Sullivan loved his men, and he wasn't going to risk endangering them in a building with a cracked wall.

After hours of effort the fire finally started to darken. The firemen had exhausted as much energy as they could muster, but each one of them knew as he stood his ground at the openings that they had this fire licked.

CHAPTER 11

Chief Sullivan's operator let the reins fall loosely between his hands. Brownie never had to be directed back to the firehouse on Bush Street for the animal had an almost supernatural instinct for the location of his stall. As they drove through the streets the chief was thinking of how this fire at Bay and Mason only underscored the potential seriousness of the city's water problem. Tonight, there had been no need to draw from the bay's endless supply of water. But what would happen if a fire broke out in the hills, most of which were far from the bay and in areas with no water cisterns built into the ground?

Dennis Sullivan was loved by his men. To them, he was not only the chief but also a genuinely great man, and San Francisco's firemen took their assessment of others as seriously as they took anything in their lives. Hailing from Florence, New Jersey, he was at fifty-three the youngest chief the city had ever had. He had just celebrated his thirteenth year in the top job three weeks earlier, having been promoted to it when he was forty. During his tenure he had made the San Francisco Fire Department one of the most modern and efficient in the world. It had the most current telegraph alarm system in the country, and the communications complex directed its force smoothly to the city's fires through 424 street alarm boxes, bells in most of the firemen's homes, and 30 bells in business premises patronized by firemen. And engine companies now had the most up-to-date steamers available, from Amoskeag, La France, and Clapp and Jones, most of them not more than eight years old and all on a rigid schedule of maintenance. Almost daily Sullivan met with representatives of firms manufacturing the new motorized fire trucks that were becoming available. He had also modernized the training battalion, and the firemen now had mandatory drill periods, with two drill towers that were convenient for their training. Fire Chief Hugh Bonner of New York, perhaps the most well-known fire chief in the world, said of Sullivan's department, ". . . [it is] as efficient and fully equipped with fire fighting appliances as any city in the world, and [it] could be learned there all that could be learned of fire departments and their methods."

Notwithstanding its reputation, the chief knew that the city, with 90 percent of its buildings constructed of wood, most of them built on uphill grades and most rows of housing originating at the mouth of a windy bay, was simply a disaster waiting in the wings. Sullivan subscribed to *Fire and Water* magazine (now *Fire Engineering*) and had carefully studied its account of the great fire in Baltimore just two years before, on February 7, 1904, that destroyed 1,343 buildings.

The Baltimore fire chief had used dynamite extensively in an attempt to control the fire by creating firebreaks, though with little success. Still, dynamite had helped contain the 1835 fire in New York and prevented it from destroying the entire city. Sullivan accordingly went to the Presidio and persuaded army officials to warehouse dynamite for the use of the fire department and to lend soldiers trained in the use of it in an emergency. There was just one problem, however: The army required a thousand dollars for the building of a dynamite shed, and the board of supervisors had denied Chief Sullivan's request for this allocation.

Baltimore had also learned from its own devastating fire that a high-pressure water system would have made a great difference in controlling the flames, a lesson that was consistent with Chief Sullivan's requests to install such a system. For each of the last six years the municipal reports had included a recommendation—a demand really—from the chief that San Francisco revitalize its water facilities. At this time, the city's water was owned by the Spring Valley Water Company, and provided primarily by a gravity-fed system from three principal locations. Thirty-three million gallons were available from Crystal Springs Lake, which fed the city via a seventeen-mile-long forty-four-inch pipe. The San Andreas Lake supplied fifteen million gallons via fourteen miles of thirty-seven-inch and forty-four-inch pipe, and thirty-one million gallons were delivered from Pilarcitos Lake through sixteen miles of pipe. In all, the city had a rated capacity of thirty-six million gallons per day, a calculation that would assure sufficient water for a city of San Francisco's size. But the three reservoirs delivered only 328 psi at the Ferry Building fire hydrant, the city's lowest level, which was simply not strong enough to feed a hose line that might have to reach as high as the eighteenth floor of the Spreckles Call Building, or any building as high in the future. The Call Building, of steel construction, had been described by the *Engineering Report* as "the best designed piece of such work in the United States." But for all that, the design of the steel would not stop a fire. Only water would do that, and water, which weighs almost eight pounds a gallon, would be best delivered from as high a height as possible and in as much volume as can be gathered in a reservoir on that high location. (The alternative was a building's own internal water supply, like the standpipe systems of New York and Chicago. Though the Palace Hotel had installed a standpipe thirty years earlier, in 1906 standpipe systems had not yet been mandated in San Francisco.) Chief Sullivan had begged the board of supervisors in the municipal reports of 1904–5 to build a specific reservoir of 10.5 million gallons in two bays on the top of Twin Peaks and to have a separate feeding system of salt water coming from the bay through two pumping stations if an emergency called for it. But still the supervisors had not acted upon his request.

Another problem was the state of San Francisco's cisterns. After a series of six major fires in 1850 and 1851, the city had installed sixty-three cisterns in strategic locations, each huge tub having a capacity from ten to thirty-five thousand gallons. But they had not been used in years, and now all but twenty-five of them were in disrepair and unusable, and those that were operable were poorly maintained by the Board of Public Works. Some had even been carelessly filled with dirt and trash by road construction crews. All clearly needed better maintenance.

The evening of the warehouse fire was unusually mild, and not a breeze was coming from the bay. The chief had always worried about the winds from San Francisco Bay whenever there was a fire. The people of the city were paying him four thousand dollars a year to anticipate the worst that could happen, and in his estimation that would be the combination of a broken water system and a wind, even a mild one. In the absence of wind, firefighters could in desperation tear a building down brick by brick if they had no access to water, but with a good breeze the fire's heat would be fanned until it slammed against every window and plank of wood, soon five buildings would be on fire, with the conflagration continuing to expand. It was his most fearful apprehension, his great black dream.

Three things are required for a fire: fuel, oxygen, and heat. (There is also a phenomenon called *halogen,* which acts as a catalyst in burning, but it does not bear on this story.) All but a few fires in America are extinguished by reducing the heat, and this is accomplished by cooling it down with water. In some fires, like those in industrial stoves, the oxygen is removed from the equation by covering the fire with foam. And, finally, in the rarest cases, the fuel is removed from the fire, usually done by dynamiting the exposed buildings that are likely to burn. The decision to dynamite buildings has never in history been taken lightly by a fire chief because his first inclination is to save property, not to destroy it.

At the beginning of the twentieth century America was a country that sustained higher than average annual fire losses than the rest of the world (and continues this appalling average today). European fire loss per capita in 1905 was thirty-three cents while the per capita loss in America was more than two dollars, or about six times greater. Americans had an adventurer's devil-may-care attitude toward fire. Like all other obstacles it was something that would be faced if it occurred. Expanding the country was what mattered to the pioneers of the nineteenth century and, unlike Europe, where buildings were constructed of thick and expensive materials like rusticated stone, marble, and alabaster, Americans built quickly and inexpensively with wood, concrete, and plaster. This lack of prudence must have been reinforced in Chief Sullivan's mind with every fire he commanded and every time he thought of the city's board of supervisors.

Why had not the city fathers acceded to Chief Sullivan's requests to revamp the water system? Could it be that they did not see the dangerous possibilities of an out-of-control fire, or was it because his department alone among all the city agencies made it a policy to list in the annual municipal reports every significant dollar allocation made to persons, firms, and corporations doing business with the fire department? Dennis Sullivan was keenly aware that the city's political powers would skin the hide off a nickel's buffalo if they could, and this annual transparency in the city report made the fire department's reputation for honesty secure and prevented, by the power of public scrutiny, the possibility of inflated charges for goods and services. Or perhaps it was because there was a significant behind-the-scenes competition among the city's businessmen to create an entirely new water delivery system for San Francisco, and the very subject of water was one to be avoided by the board of supervisors until they had made the best deal for themselves. Chief Sullivan thought their inaction to be irresponsible at the least, if not an outright malfeasance.

CHAPTER 12

California is a state of particularly staggering beauty. Within its one hundred million acres of land are enormous ranges of soaring white-tipped mountains, vast and quiet lakes, expansive and verdant valleys, easily ambling rivers and rushing rapids, forests to build a new world, and woodlands of sky-reaching redwoods unique to the world. Its almost eleven hundred miles of shoreline include natural formations, cliffs, and beaches that rival any in the world. On part of this shoreline, in the north central part of the state, lies San Francisco.

Surrounded by water on three sides, San Francisco is the 8-mile-long tip of a peninsula that runs northward for about 32 miles from Palo Alto. This wedge of land separates the Pacific Ocean from one of the largest and most natural harbors in the world: the 450 square miles of the San Francisco Bay and its 100 miles of shoreline, a geophysical wonder that gave the city its port and foundation of its commerce.

In 1906, the 46.5-square-mile city, with its population of 460,000, was the country's fourth largest metropolis and the largest west of the Mississippi. It was the showpiece of the West, and everyone who ever stepped foot there acknowledged its greatness, its vitality, its opportunity, and its surfeit of good times. It had a thriving professional class, and unions filled with skilled workers. Its strong and reliable workforce was one-third immigrant, one-third children of immigrants, and one-third more established Americans. Nearly thirty thousand Chinese provided a pool of inexpensive labor. It was a city of potential, both social and economic. But Chief Sullivan, who loved San Francisco as much as anyone, had responsibilities beyond the building of a city's reputation and financial base, and he believed the city was an investment inadequately protected from fire—something like displaying the *Mona Lisa* over a cooking stove.

In its beginning, the streets of the city were planned in a grid that would run almost north-south, but in 1836 the founding engineer, John Vioget, did not know he had to take the magnetic drift into consideration. Consequently, the grid is off north, to the northwest, by about seventeen degrees.

Whether measured by global positioning systems or magnet, however, it makes sense to think of San Francisco as a north-to-south entity, and if the city map was placed on the face of a clock with 12:00 facing north, then the Golden Gate Bridge would be at the water's edge at about 11:00 as it crosses into Marin County to the north.

To the south, eight miles from the Golden Gate Bridge, is San Mateo County, which, had the city annexed it fifty years earlier, would have given San Francisco five times as much land in which to expand, guaranteeing the future of the port and securing forever its position as the biggest city in the West.

Keeping the clock orientation in mind, meeting the water at 1:00 is Van Ness Avenue, which runs north and south. Named for a largely forgotten mid-nineteenth-century mayor, Van Ness is a wide boulevard originally designed as a firebreak. To its east is the most populated area of the city and the three most familiar of the city's forty-two hills: Nob Hill, Russian Hill, and Telegraph Hill. Other districts east of Van Ness are the Financial, Mission, North Beach, and Market Street districts. Market Street cuts diagonally across the city from about 2:00 to 5:00, and the most populated district of all was south of this diagonal line.

In the early 1900s, the city was fiercely mercantile, and its commercial hub lay east of Van Ness Avenue, or from 1:00 straight down to 5:00 on the clock: the entire business district, the great mansions and hotels, the government center, cultural palaces, Chinatown, and, South of Market, the dilapidated housing of the working poor—in all, about 80 percent of the city's value and population.

Everything west of Van Ness, called the Western Addition, makes up the other two thirds of the city and extends to the Pacific Ocean. At the turn of the twentieth century much of this area, including the expansive Golden Gate Park and the U.S. Army's Presidio, was made up of woodlands, sand dunes, and open fields and was inhabited by nearly ninety thousand people.

The hills of the city are rocky and made mostly of indurated (hardened) clay, shale, and serpentine. The lowlands are a combination of sand and clay, having been formed by runoff from the hills and deposits from the incoming tides.

No matter the substance of the soil, the city was growing, and in line with urban history it would inevitably find a way to make new construction possible. Already more than fifty high steel buildings had been built; the highest of them, at eighteen stories and the tallest building in California, was the Spreckels Building on Market Street.

Still, the city was behind the rest of the country in meeting the building reforms of the times, and Chief Sullivan was certainly aware of it. Just six months before, the National Board of Fire Underwriters had written in a report on the insurability of San Francisco:

> In view of the exceptionally large areas, great heights, numerous unprotected openings, general absence of fire breaks or stops, highly combustible nature of buildings . . . and interspersed frame buildings, the hazard is very severe.
>
> The above features combined with the almost total lack of sprinklers and absence of modern protective devices generally . . . high winds and comparatively narrow streets make the *probability* [emphasis added] feature alarmingly severe.
>
> In fact, San Francisco has violated all underwriting traditions and precedents by not burning up. That it has not already done so is largely due to the vigilance of the fire department, which cannot be relied on indefinitely to stave off the inevitable.

It was a small comfort to Chief Sullivan to know that his department was judged to be capable of doing its job. But, as every fire chief knows, there are practical and inherent limitations to what a fire department can accomplish in an emergency, which are in large part determined by the intensity of the fire, the availability of manpower, and the adequacy of the water supply. And it is always wise to remember, in anticipating and preparing for any great emergency, a fire department has another limitation—namely, that which is dictated by city politics and government support. In this respect, a fire department is essentially like an insurance policy in that the amount of life and property that can be insured is determined by the face amount of the policy: The more one spends, the more one protects. Generally, cities then as now sought to find ways to reduce the costs of fire departments, for they were, and often still are, seen as an idle, unproductive expenditure. At least that is the prevailing view until there is a catastrophe.

CHAPTER 13

REPORT OF THE STATE EARTHQUAKE INVESTIGATION COMMITTEE,

Andrew C. Larson, Chairman, 1908

Mendecino (180 miles from San Francisco)
William Mullen: The shock at Mendecino began [at 5:12 A.M.] with a tremulous motion, increasing very quickly, and decreasing also quickly. The principal disturbance was strongest toward the end. The motion seemed to be up and down, and also from north to south. . . . It lasted about forty seconds. Beds were moved from three to five feet, and pianos to the same extent. . . . A rumbling sound, like distant thunder, preceded the shake, and was loudest at the commencement of the movement. During the shake, animals became greatly excited. . . . Water in some wells became muddy and frothy.

Clear Lake (120 miles from San Francisco)
Charles M. Hammond: My wife and I . . . were awakened by a violent rocking of the house. We jumped to a doorway and stood there for about two minutes, the house gradually coming to a state of rest from its violent rocking and swaying, and a roaring noise passing off in a southwest direction.

Point Arena (95 miles from San Francisco)
W. W. Fairbanks: A heavy roaring sound preceded the shock. The ground moved in undulating swells or waves, rising and falling. Men and animals—horses, cows, etc.—were thrown to the ground, and were unable to rise or stand during the

shock. . . . All brick buildings were thrown to the ground. . . . In many cases houses drifted away and left porches standing in their old location.

Santa Rosa (55 miles from San Francisco)
J. W. Brown: On going outside [I] heard a great noise from the west and saw the treetops waving. The noise and motion of trees approached [me], and [I] took hold of a small tree nearby for support. This tree was torn from [my] grasp. The ground seemed to be in waves about two feet high and fifteen feet long.

CHAPTER 14

At 5:12 A.M., Wednesday, April 18, the magnificent dome of San Francisco's city hall began to shake. Built by the corrupt political boss Chris Buckley for more than $7 million in the 1890s, during a period when Boss Tweed built his famous courthouse in New York for $13 million (about $175 million and $325 million, respectively, in today's dollar value), it had a classical elegance and beauty that were much admired. But now its concrete base, its marble columns, and its symbolic statuary tumbled directly onto the pavement of McAllister Street, killing several people. In the hall's basement hospital, jail, and insane ward others scrambled for their lives.

Rudolph Spreckels, just sixteen blocks north on Pacific Street, did not hear it fall, but something stirred him from the dark of his sleep. A second went by before his eyes opened and he realized what was happening. The sound began as a distant thunder growling in the earth and running through the house, and as Rudolph rose on his elbows, it grew into the churning of a heavy locomotive speeding toward him, becoming louder and louder, until, at the fifth or sixth second, the movement started. He reached out for Nell's hand as, suddenly, his room seemed to bounce, as if it had been struck by a meteor. The floor was struck again and again, an angry intermittent jolting, and began oscillating up and down. Nell was awake now, and he enveloped her in his arms. He had to do something, to get her to some protection—the doorway perhaps, where there might be some support if the house collapsed. The floor was rising and falling about a third of an inch and as much as three quarters of an inch with each new vibration, maybe four times per second, continuously for forty-five unrelenting seconds, each jolt a shock of fear that the world itself was falling apart. He pulled Nell from the bed and, barely able to remain on their feet, they moved to the door of the bathroom, for the transom was nearer to the street there. Everything loose danced and chattered, and he could see the marble on top of the bureau across the room tapping furiously, and the chandelier tinkling like door chimes in a small cloud of plaster dust. The vibration increased in power as the seconds wore on, and each vibration seemed to flow through his body as if someone were pummeling him. He held his wife tighter, her head cupped into his shoulder. A straight chair in the corner and then a thin bottle of lilac lotion on top of the bureau marble tipped over. A small picture of a Hawaiian palace fell from the wall, breaking the frame's glass. Outside, the gate and iron fencing around the house toppled, and the concrete walkway before it cracked.

Throughout the city buildings that had been constructed on the hard clay of high ground were safer than those in the low waterfront areas of sandy and filled-in soil, where they began to sink into the liquefying earth, first one story disappearing and then two, dropping suddenly into an earth that no longer had substance. Other buildings threw off their pediments and cornices while their parapets—those short walls facing the edges meant by code to protect firefighters—simply fell into the street. Still other structures collapsed entirely, falling inward or outward, depending, like the weakest link in a chain, on which part of the construction gave out first. Chimneys fell everywhere, and a pall of plaster and cement dust filled the early-morning sky. In the harbor, two derricks toppled and fell through the ship *City of Pueblo,* cutting her cleanly in half. Five hundred monuments keeled over in the city's graveyard, all toward the east and northeast, and in the Laurel Hill Cemetery two similar granite shafts two hundred feet apart were turned in opposite directions.

For miles in every direction a harshly waving earth caused stress in the foundations of countless buildings, many of which were unable to withstand the undermining of their stability. Up and down the Pacific coast in idyllic towns with romantic names like San Jose, Santa Rosa, Palo Alto, Cloverdale, and Santa Cruz houses shook and fell, reduced to ash and brick piles. The shaking came from the west, eight miles out to sea from San Francisco, and caused destruction as far as fifty miles to the east and two hundred miles north to south, from Monterey to Humboldt County.

At Point Arena, ninety-five miles north of the city, a sixty-foot lighthouse came crashing to the ground, and ninety miles north of that a ship called *Argo* jolted and jumped in the water so abruptly that the captain believed he had rammed another vessel. A bridge at Alder Creek, famous to all Californians, toppled into the water. The entire town of Fort Bragg was leveled, and at Point Reyes the 5:15 train to San Francisco, the huge locomotive and its half dozen cars, was thrown over on its side next to the tracks.

At about the twenty-fifth second after the rumbling started, Rudolph began to feel a reciprocating jerking added to the vertical lurching, so that the world now seemed to be moving both up and down and back and forth. This movement increased in amplitude until he felt a sensation like spinning on a wave, for there seemed to be no middle anywhere, no center of gravity that could be recognized; that certain center of gravity that he had unthinkingly intuited in every step he had ever taken was now utterly gone.

Consider a weight on a string, swinging back and forth like a pendulum. If the weight is hit by a force at the lowest point in its path to the earth, that action will create another pendulum going in the direction opposite the force, due to the simple Newtonian explanation that for every action there is an equal reaction. If the force, however, hits the weight at anywhere other than at the lowest point in its path it will not create another pendulum but a circular motion. This circular action combined with both the pendulum motion and an up-and-down motion are the three most fundamental movements in nature. If oscillation and waving are added to these three motions, the net effect will approximate the movement of the ground that might be felt by a person during an earthquake. It can be beyond Newtonian in relation to the original point of the quake, and if one is inside a house, one will move the way the house moves, and the house will move according to its construction and the soil it sits on.

There were four classes of buildings in San Francisco. The first class was the government structures. Those built by federal funds and according to federal standards were strongest while the city and state buildings, which were more often subject to limited funds or compromised by political or labor corruption, were less durable. Then there were the oldest structures, built of wood. Ninety percent of the city's buildings were made of wood, and many of these had been constructed on "made" ground—the shoreline on the bay side, swamps, creeks, and wetlands that had been filled in. Those that were built on wooden pilings driven deep into the land and tied to the foundation were relatively stable, but most had been built on the raft principle with a flat foundation of bare wood. Then there were buildings of unreinforced concrete that, if constructed with the high quality of Portland cement, wetted brick, and strong joists and anchorage, were much stronger than those made of timber and exterior brick laid with common mortar, unless they were built on very firm ground. And, finally, there were the high steel-frame buildings that, because of the demand of their owners for more light and glass, were less strong and resilient than those that liberally used wind-breaking diagonal beams that reduced swaying.

As Rudolph and Nell huddled in the doorway, he began to consider the solidity of his home's construction. Unlike Henry Dutton's expansive Italianate Victorian mansion just across the street at 1782 Pacific Street, with its jigsawed ornamentations, a rich man's prize in blue and yellow, the Spreckels house was classical and conservative, made of stone and very large, with a circular drive bordered by trees. It looked Palladian, with four two-story ionic pillars protecting its five entry steps, its perfect balance broken only by an attached carriage house on the left, or west, side of the house. It also rested on deep pilings set into the hardened clay of the city's hills, which would allow it to sway with the movement of the ground.

The house would be all right, he was thinking. But Nell? How long would this damned shaking last? He knew he had to get her to the open air, to safety. And the children, too. The servants? Where were they and were they safe? He began to run for his trousers and for a covering for Nell.

Suddenly the jolting seemed to get worse, if that was possible, the shocks growing even more violent. They began to come in waves of three pulsations, with a power and battering force that had never before been felt in San Francisco, and in just a few places in all of history. Rudolph now believed the end of life was imminent, as indeed did so many Californians. He would have to keep Nell close to him if the house caved, and perhaps he might protect her and cushion her fall.

The shaking continued, lasting thirty-five seconds, and then forty, and then finally, at about forty-five seconds, it began to ebb. At forty-seven seconds it was over. It was at that moment that the huge and Florentine dome of city hall, the symbol of San Francisco's strength, wealth, and sophistication, slowly crumbled to pieces.

CHAPTER 15

Lieutenant Freeman felt the sudden jolt, as if someone had lifted the *Perry* and slammed her into the water at the pier's edge, leaving the entire boat vibrating. But it was a strange vibration, one unknown to most seamen, for it was vertical in orientation. Waiting to see if it would pass, Freeman looked around his small stateroom as he tried to jump from his cot. He saw Midshipman Pond, in his slippers and robe, holding himself upright by bracing himself in the threshold. Both had spent enough time in earthquake areas—Freeman in California and Pond in Hawaii—to be familiar with how the impact of a seismic disturbance is transmitted through the water. Water does not shake as the land does, but it does have a serious if momentary displacement before it seeks its own level. Indeed, it has been reported that entire swimming pools have been emptied of water in a single earthquake jolt.

The two men traded a glance, agreeing that they were experiencing an earthquake, and a bigger earthquake than they had ever conceived of. Freeman put on his own dressing gown and shoes, and they made their way to the deck, where, looking out at the water, they were immediately struck by how choppy it was on such a windless morning. It would have been clearer to them what was happening if they could have seen, as was later reported, a Native American leaning over at the bay's shore not far away in Marin County. The Indian was scraping mussels from a rock grouping and was amazed in the shaking to see the water level go quickly down, giving him a chance to scrape off such an unusual crop of mussels before the water rose again a few minutes later.

CHAPTER 16

Donaldina Cameron lost no time as soon as she felt the first tremors. Earthquakes were no stranger to her for she had been living in San Francisco since 1895. A significant quake had struck in 1898, which was severe in Oakland, but only forceful enough in San Francisco for people to realize that the earth was moving. And so Donaldina recognized the danger and was certain of her responsibilities. She threw her robe about her as she charged down the hall to the girls' dormitories. She could hear the little ones crying and saw the terror in their eyes. "It is all right," she assured them as she ran through the room, trying to calm them. "It is all right." The older ones immediately began to care for the children, and in the midst of a trembling earth Donaldina was moved by the absolute lack of cowardice among her girls.

A chimney had fallen to the roof of 920 Sacramento Street, and plaster dust and bric-a-brac had begun falling all through the home. Donaldina mustered her girls and staff together and they waited along the interior hall for the shaking to stop.

Yuen Kum, at twenty one of the oldest in the shelter, had each arm over the shoulder of a little girl, hugging them for comfort. More than fifty of them huddled together in the hallway. It was so frightening for the children, Yuen Kum was thinking. If only Henry Lai was here. Where would Henry be now? How will God see that he arrives in this calamity?

CHAPTER 17

Just three blocks away, in a mission house of the Salvation Army at Sacramento and Grant, Ensign Susie World, a young soldier in the Christian army, barely thirty, was awakened by the screaming of her lieutenant and by the crashing down of part of the ceiling in her room. She then heard the cries of her visiting niece and nephew, "Auntie, Auntie . . ." With the entire house reeling and swaying, Susie ran to the lieutenant, who had hurried to the front of the mission, prepared to jump with her baby in her arms from the second-floor window. Susie grabbed her shoulders and pulled her toward the middle of the room, thinking that it might be safer for them there. Suddenly, the shaking became more violent, and as she recalled, "There seemed to be some gigantic, unseen force trying to tear everything to pieces. Then, with a terrible crash, our front wall fell to the street. Stunned, we waited for the next crash that we thought might usher us into eternity."

Seeing to the street through the huge holes that had been left by the collapse of the walls, Susie determined that they should immediately head outdoors. She began to guide her charges down the stairs and was climbing over the fallen walls when she realized that she was wearing only her nightshirt, and her lieutenant was barely covered by a silken sleeping gown. It was embarrassing to be about in the world so immodestly, but she said a small prayer and held tightly to the hands of her young niece and nephew. The earth was still churning, and as they began to run over the rubble, Susie lost the slippers she had stepped into. The children were barefoot as well, but made no complaint. They sought refuge beneath the arched entryway of a steel-braced building, and from there Susie saw six large dray horses lying on the ground, all electrocuted from a wire that was still arcing. The small group clung to one another to share warmth and comfort. Just next to them was a dead man, crushed, apparently, by a fallen chimney. People were now running past them, and ambulances, patrol cars, and horses were speeding by with all manner of carriages and charges. Susie had no idea what she would do next. In the rising light of early morn, it was still dark and chilly, and she could feel her niece trembling. Her nephew had pulled along a blanket from his bed, and using a fallen stone Susie ripped it in two, wrapping the other half around her niece's shoulders. Susie, too, felt the cold, and again was ashamed by how little she was wearing. Then a woman appeared, and like a gift from the Almighty, she presented two skirts she had taken from a shop next door. Susie and her lieutenant quickly tied the skirts around their waists, and this saving grace gave them the courage to consider what they would do next.

CHAPTER 18

Six miles from San Jose and about fifty miles from San Francisco was an asylum for the insane at Agnew. Along with San Francisco's City Hall, it would become notorious after the quake as a leading example of the poor construction engineering, shoddy workmanship, and inferior building materials of the time. When the shock waves hit the central tower of its main building, the stone and concrete simply fell apart, crashing down on the fleeing patients and administrators, crushing 117 of them to death. Several hundred patients ran terrified around the grounds, wide eyed and frantic, until they were apprehended, many of them then tied to trees until the situation could be brought under control by the authorities. History would judge the incident as pathetic and tragic while the contemporary newspapers reported it as comic relief in the midst of drama.

CHAPTER 19

Jack Murray had barely rested that night, the smoke from the third-alarm fire still in his hair and his nostrils. But then his muscles suddenly ached the way they did whenever he was interrupted from sleep. The firehouse was a solid stone building, but it shook the way the ground shook, and the ground seemed to be pounding upward. Jack, in all the years he had lived in San Francisco, had experienced dozens of earthquakes. But this was nothing like he remembered. He thought only of Annie and the children as he clambered out of bed, trying to stabilize his footing. The family was on high ground, he thought. Telegraph Hill was solid rock. The building would stand. They would be all right. But if only he could be certain.

Captain Murphy was in the bunk room before the shaking stopped, calling for his men to protect themselves. The beds were clanging on the floor and the springs were singing. Even as the building shook, the firemen slid down the pole, and headed for the street. Jack was driving, and so he went to the horses, which were rearing and neighing and stamping their feet in terror. This is some night, he must have been thinking—a third alarm, and now this.

CHAPTER 20

Dennis Sullivan was in his third hour of sleep after the Bay Street Cannery fire in his bedroom on the third floor of the firehouse of Chemical 3 at 410 Bush Street, right off Kearney (now spelled Kearny). He immediately rose in his bed at the first roar and rumbling, certain that it was an earthquake, for nothing else has that distinctive sound. He assumed it was probably one of the many small tremors that San Francisco experiences every year, but soon noticed that this roar was more substantial than any he had ever before heard. He felt his bed shaking, and as it quickly began to convulse, all of the responsibilities and apprehensions of his rank and office began racing before his eyes.

Most people in such a situation would rightly be concerned for their own safety, but Dennis Sullivan's principles and the hundreds of emergencies he had attended taught him to look out for the welfare of others in any abnormal or crisis situation. Now, though, his first thought was for the horses and how the firemen had to get into their stables to calm them as they would almost certainly be needed to pull the steam engines in the hours ahead.

And then Maggie overtook his thoughts, Maggie who was asleep in the adjoining room. "Margaret," he yelled, and just as he threw himself over the side of the bed and began to find his legs in the terrible shaking, the sound of a loud and fierce crash filled the firehouse. It stunned and shook him further, knocking him off his feet and leaving him flailing. The chief rose to his hands and knees and struggled to get up as the floor continued to wave and pound. "Margaret," he yelled again, and finally managed to run into the adjoining room, where he encountered a blinding cloud of dust. He did not realize that the thin cupola of the California Hotel, which stood next door, had toppled in the shaking and penetrated the firehouse roof, piercing the floors beneath like a knife as it fell. Maggie, tucked firmly into her bed, dropped in the path of the steeple down to the second floor, and then to the first, and finally through the apparatus floor to the basement. Because of the dust the chief did not see the hole in the floor on the third floor and stumbled into it, plunging to the basement himself.

Maggie, still tightly bound by her sheets, had fallen more than forty feet, and miraculously her bed landed upright. Still, the concrete and plaster continued to fall, covering her completely so that she could barely be seen beneath the rubble.

The chief, though, came to rest near the rear of the basement, against the heating furnace, a large coal-fed boiler connected to the water and steam heat system. One of its galvanized pipes had broken apart and was gushing steam with the force of a large hose. Dennis Sullivan was lying in the immediate path of this thick flow of boiling hot water and air.

The chief struggled in the extraordinary pain. He knew he had to get away from the pipe, but he was covered with rubble and had many broken bones. He struggled with all his strength to move the pile of wood and concrete that braced him.

Chief Walter Cook had been asleep on the second floor of the Bush Street firehouse along with the company members of Chemical 3 when the quake began. They jumped out of their beds on hearing its roar, and Chief Cook saw out of the corner of his eye a fireman named John Coyne losing his balance at the precipice of the hole in the floor. He ran to steady him, but was too late, and Coyne disappeared through the hole to the floor below, landing near Mrs. Sullivan on a coal pile in the cellar.

The firefighters ran down the steel steps and quickly began to dig through the rubble for the chief and his wife, both of whom they could hear moaning and begging for mercy. Coyne had picked himself up and, except for cuts and bruises, was unhurt. Newspapermen from the *Bulletin* office across the street hurried over and joined several police officers now working alongside the firefighters to dig Mrs. Sullivan out. They created a brigade line to carry out the bigger pieces, and just as they uncovered her face, the fire chief himself, like a surprising apparition, climbed out of the shadows at the rear of the firehouse. Several firemen ran to him and caught him just as he collapsed. Since the firehouse's horses were either dead or helplessly trapped behind the rubble, they carried him quickly to the St. George Stables, just next door, and asked the proprietor, John Murphy, to harness a rig to ambulance the chief. "Maggie," Dennis Sullivan was crying, "we can't leave without Maggie."

"Easy now, Chief," they said, lifting him onto a wagon. They drove him to the receiving hospital, which was located in the basement of city hall, but it was obvious that the hospital could not function in the destruction of the mammoth building. Their next stop was at the Southern Pacific Hospital at 14th and Mission, sixteen blocks from the firehouse.

It took twenty minutes to free Maggie Sullivan. Chief Cook supervised the firemen as they carefully lifted her from the rubble, and placed her on a makeshift stretcher of blankets. They carried her to the lobby of the California Hotel, where a doctor took charge of her.

A clerk then yelled to Chief Cook that a Dr. Stinson was buried up on the eighth floor by the bricks that had fallen when the cupola tipped over and went through the firehouse. John Coyne was limping, but he ran with the chief up the eight flights of stairs. They got on their hands and knees and began pulling aside the debris until they found the doctor, who had been crushed by the bricks. It was the first death of so many Chief Cook would see this day.

Walter Cook looked out of a window from this high vantage point over the city and what he saw was astounding. Fires were raging in every direction, all burning freely, fifty alarms reported in that first half hour alone, caused by overturned candles, heaters, broken flue pipes, and ashes spread from toppled cooking stoves. Only one thought must have run through his mind: San Francisco is burning.

He was also thinking as he surveyed this horrific scene what every firefighting specialist in the city would conclude: that it might be one of the greatest tragedies to befall the city that Dennis Sullivan would not be responding to the biggest alarm of his tenure. In his absence, it would become the responsibility of less qualified men to try to fill the firefighting void, in the process making decisions that would come to destroy the city, place thousands of lives in jeopardy, and force hundreds of thousands from their homes.

CHAPTER 21

At 676 Howard Street, nine blocks from the California Hotel, James O'Neil of Truck 1 was walking toward the back end of the Engine 4 firehouse to the stalls. Rule 26 of the department regulations required that the horses be watered at 5:00 A.M. and be fed at 6:00. Jimmy realized he was ten or fifteen minutes late in watering the eight horses that belonged to Engine 4 and Truck 1, but the horses of the fire department were well known for being uncomplaining.

There was still something of a pall over the firehouse since the line-of-duty death of Engine 4's hero captain, Charlie Dakin, a few weeks before, but the men were coming around, adjusting to the terrible shock. The department day started when Jimmy would tap the house bells at 6 A.M. and the men would amble down to the apparatus floor. One of them would begin cooking some oatmeal, another would wash the stalls down with boiling water, and still another would buff the harnesses. The everyday cleaning would begin, the beds would be made, the floors swabbed, the tools—axes, crowbars, pulling hooks—would be wiped clean and oiled, the poles would be rubbed with brass polish until they gleamed.

It would be a normal day, O'Neil thought, until suddenly the ground shook and the southern wall of the American Hotel, just next to the firehouse, fell away from the body of the structure and crashed into the firehouse. It separated quickly and quietly, giving no warning, the way a board toppling from a roof construction would have. Had Jimmy heard the rumble he would have run for the safety of a threshold, but instead he simply stopped to gain a momentary confidence in his footing. But then the redbrick wall just next to him was pushed in by the collapse of the neighboring hotel. James O'Neil did not draw another breath; as the bricks pushed him into the concrete floor, he died instantly. Across the street from the firehouse, the floor collapsed in a Chinese laundry, and the fires used to heat the many irons tumbled across the floor and the building went up in flames. The Chinese laundry fire would grow to be one of the worst spreading fires in the city.

O'Neil was the first of the two firefighters who would lose their lives to the earthquake. The second, the great Fire Chief Dennis Sullivan, succumbed four days later from injuries sustained by the falling cupola of the California Hotel. In all that time, he lay dying in his hospital bed, unconscious for the most part, and unable to sustain the pain and severity of the burns that would finally take his life.

To the end, the chief was never aware that he was a victim of the greatest city fire in peacetime history or that he had been unable to take part in the biggest alarm of his career. He could take no pride in the fact that the fifty-two fires reported in the first minutes after the earthquake were fought aggressively and courageously by his firefighters, that his legacy of training, training, and more training was instrumental in the containing of those early blazes. His men had isolated almost all of those fires, and none of them would have gotten out of control but for misguided decisions made by the city's leadership. All of them would have been contained had Chief Sullivan been able to execute for the city the authority of command for which he had devoted his life's experience and education. It was the single biggest tragedy of April 18 that Dennis Sullivan was kept by providence from trusting in his own confidence to make legally constituted decisions, decisions that would have, in all probability, saved the City of San Francisco.

CHAPTER 22

The men of the slaughterhouse district started their work at 4:00 A.M. each morning. With little refrigeration available beyond iceboxes, the abattoirs—companies like Miller & Lux, Jacob Schoenfelt, and Poly, Heilbron & Co.—had to have their meat, chicken, and fish delivered to the hotels and restaurants of the city before the high sun, so that the food would not begin to turn. This busy district was located at the water's edge where 3rd Street meets the bay, then near Townsend Street. There a stable, one of the largest in the city, held exactly two hundred horses used to pull the meat wagons. They had not yet been hitched, and the animals, some of the men noticed, shuffled their feet moments before the earthquake, until in the high volume of a strange chorus they began to whinny as the ground beneath them began to thunder. The stable floor was built on pilings set into the bay-shore water, and as the ground continued to shake without relenting, the pilings fell over, sliding the whole of the stable into the water. The roof also collapsed, but landed on the fencing between the stalls. The desperate animals were forced to their knees in the most unnatural positions and then were pushed over onto their sides. Many were bleeding; two had broken their legs. They began snorting in a weird cacophony, eructing an anxious apprehension, their eyes wide with fear of the unknown. Their withers quivered, and their legs struck out against the floor and the sidewalls. And then the tide began to come in, leaving them helpless as the water began to rise.

Several hundred men rushed to the waterside, carrying tools, crowbars, screwdrivers, and hammers, and waded through the rippling waves until they reached the horses. The roof could be dismantled, they realized, but how would they lift the horses out? The most they could hope for was to drag the poor creatures out of their stalls, and so they worked furiously to remove the siding of the building and to cut away whatever risers they could without further endangering the animals. As they pulled free the wood the animals grew quiet, as if in recognition of the urgency of the frenzied activity around them. But as the tide was coming in quickly, a man crawled into each stall to soothe the horses with low, reassuring tones, holding their heads high enough so that the water would be out of their nostrils. The men waited patiently until the walls were torn free and the early-morning light shot through the confining stalls. All of the horses eventually survived the collapse, except for the two whose legs had been broken. Their rescue was one of the many striking incidents in the days that followed of human beings coming together in common purpose in the midst of a catastrophe.

CHAPTER 23

Just three blocks south, at 6th and Howard, in a dried-up pond that had caved into the ground in a previous earthquake, a series of buildings had been constructed on filled earth. As the rumbling and vibrating began, the Nevada House Hotel began to tilt over and then collapse fully on its neighbor on 6th Street, the Lormor Hotel. The Lormor in turn fell against the Ohio House Hotel and finally, the weight of these three buildings tumbling like dominoes hit the Brunswick House Hotel, which was then pushed over and far out into Howard, as far as the rail tracks running up the middle of the street. Hundreds of people were trapped in the more than one thousand rooms of these hotels and, almost immediately, fire was seen burning at the edges of the pile. No firemen were available as all of them in this area were already operating at the Chinese laundry fire. Screams were heard all about, the mournful cries of agony of those trapped beneath tons of wood. These were the poor of San Francisco, the unregistered and undocumented transients living in the cheap hotels and rooming houses south of Market Street. Two police officers immediately climbed to the top of the rubble heap and began to pull out anyone they could. Two or three people were saved, but the fire spread quickly and was soon roaring. Several men held blankets between the policemen and the flames so that they could continue to dig as long as possible, but they were finally forced to withdraw. The flames passed over the voices until the voices were heard no more.

The hotel blaze became the murderous fire that would eventually travel across Minna and Mission and Jessie streets and travel north up Market to the Spreckels family high-rise, the tallest building in the West.

CHAPTER 24

On Mission Street, a group of cowboys was running fifty or sixty long-horned steers from the waterfront to the Potrero stockyards when the shaking began. The steers bolted, wild eyed and desperate, just as police officer Harry Walsh stopped near Fremont Street to speak with John Moller, the owner of a saloon at that location. Just after gaining his footing on the moving ground, Walsh saw the herd of cattle racing toward him, and he drew his firearm. He knew that the small revolver would do little to stop so large an animal unless he managed to center his bullet in its forehead. But as he lifted his gun, a wall fell into the middle of the intersection, covering many of the frightened beasts and killing them instantly. A few, though, were trapped beneath the piles of brick and whining in pain. Walsh ran to them and shot two to end their misery but then saw another group of steers coming toward him. He pulled Moller back, and they both began to run toward Moller's saloon. Moller stopped without explanation and put his hand out as if to stop the thousand-pound creatures. With the few bullets he had left, Walsh emptied his chamber into the onrushing steers, but it was too late for Moller, who had been hit and trampled. Walsh borrowed a rifle from someone in the gathering crowd and began bringing down the panicked steers as his friend Moller was rushed to a hospital. But he was relocated twice in the next five hours after forced evacuations, and Moller, like many that day who were to be shuffled from hospital to hospital, was unable to sustain his traumas. He died, finally, at the close of the morning.

CHAPTER 25

In earth science, it can be said that the only constant is that there is nothing constant. The earth is always changing, every second of every minute, both on the surface and beneath the earth, thirty-three hundred miles down to the edge of the core, where the temperature is estimated to be as hot as the surface of the sun. There these temperatures cause continual pressure—pressure so great it can shift the skin of the earth far above. This shifting is what we call an earthquake, an event that takes place some eight hundred thousand times a year.

The word *great* in its most substantive meaning can be applied to the consequences of this subterranean stress. As Newton proved, for every action there is an opposite reaction, and when pressure builds within the earth it eventually finds a way to seek release, either in a volcanic eruption or in an earthquake. Great things happen as a result—great continents, great mountain ranges, and great rivers can be created, great numbers of people can be killed.

Geologists, seismologists (*seismos* is the Greek for *earthquake*), paleoseismologists, and physicists have come to learn a great deal about earthquakes and volcanoes. But our knowledge continues to be very much wanting, for we have been constrained in our observations by time. We, as Homo sapiens, have been on the earth for just a metaphorical minute in relation to the amount of time the planet itself has existed—perhaps two hundred thousand years in relation to four and a half *billion* years. During this time there have been tens of billions of earthquakes of varying degrees of severity and almost that many tsunamis.

Disaster is a word that we humans attach to an event to give it status or simply nomenclature, but for the earth an earthquake is no more disastrous than the opening of a lilac's petals. It is part of the natural order, like wildfires, tsunamis, landslides, avalanches, floods, droughts, blizzards, cyclones, tornadoes, and lightning, and becomes catastrophic only when human beings are involved.

Throughout geologic time the earth has had many different configurations of land, water, and inhabitants. Certainly, it has been subjected to many upheavals, and whatever landmasses have existed must have shifted in the course of the planet's history. About 250 million years ago the earth was configured in a single great mass of land scientists call the Pangaea, and the rest of the planet, for the most part, was covered by a body of water known as the Tethys Sea. Some unknown event then occurred that initiated a process that caused the earth's plates to move in what we now call continental drift.

Indeed, it was Sir Francis Bacon who first noticed at the end of the sixteenth century that "the coast lines of South America and Africa would fit together perfectly if the ocean were not between them." Bacon, one of the more strikingly distinguished adventurers in history, wrote no more about the subject, leaving his observation of continental fit as a task for future thinkers. It took more than three hundred years for someone to accept the challenge.

In 1915, Alfred Wegener, a German meteorologist, published "The Origins of Continents and Oceans," a long essay in which he offered his theory of plate tectonics, or the movement of landmasses—a theory that was largely dismissed for the next twenty-five years and hotly debated until recently. In understanding plate tectonics it is important to distinguish between the earth's chemical subdivisions (the core, mantle, and crust) and the two mechanical layers nearest the surface—the lithosphere, which is relatively rigid and cool rock, and, about fifteen or twenty miles farther down, the asthenosphere, which is much hotter, sometimes molten, and made of mechanically less substantive rock material. Together the lithosphere and asthenosphere are about sixty miles deep, but where one ends and the other begins is hard to discern. The thinner but stronger lithosphere consists of about nine large plates and about twenty smaller ones (the precise number of plates is not known), each separate and distinctively identified by plate boundaries. On these plates float the one-third land and two-thirds water of the earth's surface.

These plates are so tightly bound together that they cannot move without a great force. The incalculably vast amount of heat below them in the asthenosphere causes convection currents, which create pressure and stress on the plates. This stress builds up for years, sometimes centuries, until the plates can no longer stay joined together. The accumulated pressure causes a plate or plates to move, and it is this movement, usually along a weak section of a plate called a fault, that gives rise to earthquakes.

This movement can occur in three ways. The first is called slippage, or a transform fault, and takes place when plates slide horizontally against each other. This is how the San Andreas fault moves—at least, we can determine that its slippage has been horizontal for the last million years. Still, we really don't know for certain if it has always been and always will be a slip fault.

It is simple to envision this movement if you press your hands together as tightly as possible and, then, with a sudden jerk, move your right hand forward about an inch. Now think of each of your hands as being a plate about fifteen miles in thickness, and you can understand the awesome power of this movement as it sends shock waves in primary and secondary waves to the surface of the earth.

It is these waves that do much of the damage in an earthquake. To get a sense of the power that results from the release of so much pressure and stress, bend a yardstick to near its breaking point, then release one end. You can see how forcefully the stick shakes. Now imagine this shaking accompanied by waves consisting of millions of pounds per square inch, traveling from fourteen thousand to twenty-eight thousand miles per hour, or four to eight miles per second.

The second movement is called a divergent movement, or a rift, where plates are being pulled apart. This is the kind of movement that extends the ocean floor and separates continents. If you took two yardsticks glued end to end with the strongest glue available and pulled them apart, using as many people as needed to pull each end until the yardsticks separated, and if you thought of the yardstick as being not a quarter inch in thickness but fifteen miles, you can visualize the dynamics of a rift quake.

The third plate movement is called a thrust, or convergent, which is when one plate pushes beneath another. The faults near Los Angeles are thrusting faults, as are those in Alaska. Normally, these occur when the heavier basaltic crust of the ocean pushes beneath the lighter granite crust of a continent. This type of fault has created an abyss in the sea floor near the coast of Japan that descends more than seven miles, the greatest ocean depth known. When one plate sinks under another it is theorized that it meshes with the magma beneath and consequently loses its solidity. Thrusting faults are the most fearsome of the three, but that is not to say that one type of earthquake will inevitably do more damage than another. The degree of destructiveness depends on what is built on the surface in proximity to the epicenter and how many people are present.

When we think of the San Andreas fault as two massive plates, one under the Pacific and the other under the continental United States, slipping alongside each other for a length of eight hundred miles and a depth of ten miles, we can begin to imagine the potential power generated by this huge amount of earth material. Such plate movements occurred throughout the world in unusually high numbers in 1905 and 1906. Just the previous September, an estimated 7.9 earthquake shook Calabria, Italy, killing 2,500. On January 31, an estimated 8.8 earthquake struck Ecuador, 120 miles west of Tortuga, with a resulting tsunami that killed more than 1,000. It took the waves twelve and a half hours to reach Hawaii, and when they arrived they caused much destruction to villages on the bay at Hilo. In February a series of earthquakes shook the island of St. Lucia. One of them was estimated at 7.0 and was felt 120 miles away in Grenada. On March 16, a 7.1 earthquake on the island of Formosa killed more than 1,300 people. On April 6, only twelve days before the San Francisco disaster, Mount Vesuvius erupted in an explosion that ripped 325 feet off its top and that blew ash for ten days, killing more than 2,000. And the following August, an 8.6 earthquake hit Chile, one of the most seismically active areas in the world, which killed more than 1,500.

For the Chinese, 1906 was the Year of the Fire Horse, an event that occurs every sixty years, an apt description for a year with so many earth movements and so much loss of life.

The San Francisco earthquake of April 18 is estimated to have been of 7.8 magnitude, a level that no quake in the continental United States has reached since then. However, seven earthquakes in Alaska have been of greater magnitude and one in Hawaii. Indeed, in 1964, an earthquake in Prince William, Alaska, reached a magnitude of 9.2.

In understanding earthquakes it is important to distinguish between measurements of their intensity and their magnitude.

In 1906, earthquakes were measured by various scales of intensity, such as that found in the Rossi–Forel Scale, which was developed in the Alps as a way of appraising the degree of local damage during and after a quake. Until then earthquakes had been scaled by the amount of lives lost and injured, building and road damage, and earth displacement. But none of these measurements provided viable correlations that could lead one to understand precisely what happened when the earth heaved and so could not offer an accurate scientific assessment of the event. In 1906, for instance, the earthquake was felt in Sacramento but did hardly any damage there, which meant that its intensity varied widely from place to place. Also, it was reported that bricks were thrown off buildings across the street from the city's Ferry Building during the quake, data that were incorporated into a Rossi–Forel Scale assignment. In fact, it was discovered later that the bricks fell when fire weakened the wooden joists of the affected buildings and not during the quake itself, which rendered the measurement inaccurate.

But the 1906 earthquake was also recorded by seismographic instruments and in places as far away as Europe. Their measurements took account of the total energy of the earthquake, or the amount of energy expended from its epicenter and outward in all directions.

In 1935, Charles Richter of the California Institute of Technology recognized the need to measure magnitude precisely and so constructed a logarithmic relationship between ground motion and the distance from an epicenter, based on the wavelengths measured at a number of seismograph sites. He correlated his estimates with specific numbers, so that a 3 to 4 reading was often felt but did not cause damage, a 5 was widely felt with slight damage near the epicenter, a 6 did damage to poorly constructed buildings within 10 kilometers (6.2 miles) of the epicenter, a 7 was a major earthquake with serious damage up to 100 kilometers (62 miles) from the epicenter, and an 8 was a major earthquake, like the 1906 shaking in San Francisco, which featured great destruction and much loss of life over several hundred miles. The Richter Scale 9 would result in significantly major damage over a distance of more than 1,000 kilometers (620 miles), but one had never been recorded.

In 1978, Hiroo Kanamori, also of Caltech, recognized the inadequacy of the Richter Scale. He saw that the larger the earthquake, the less accurate its reading on a seismograph. Kanamori then developed the Moment Magnitude (MM) Scale, which is a measurement of the total energy that is released by an earthquake. It is similar to the Richter Scale and uses a comparable numbering system, but it is more precise, particularly for more severe earthquakes. (For example, the Richter Scale probability, a historical estimate, for 1906 would be an 8.3 level earthquake, while an

MM Scale would have been 7.7.) The Moment Magnitude Scale is the measure used predominantly by scientists today, even though the press continues to report earthquakes according to the Richter number.

Plates, however, do not need an earthquake to move. We know that Baja California is moving away from Mexico in a northwestward direction at the rate of two inches a year and has been moving for four to six million years. We know that Europe is moving away from the United States at the rate of one inch every year, and that Maui is moving away from South America at the rate of three inches a year. Geodesy is the science of the shape of the earth, and with the advent in the last decade of global positioning systems, geodesists in the future will be able to map, from the readings forwarded from a space satellite, every movement of the land and sea with authority and exactitude.

Plates have been theorized to have always been in movement. Indeed, it may be that even before the creation of Pangaea 250 million years ago, there had been many other landmass configurations, being shaped by the haphazard movement of plates. Scientists are not sure. There is much that we do not know, and even the U.S. Geological Survey (USGS) admits that there are certain things—like the pre-Pangaea shape of the land and what keeps the core of the earth hot—that we will never know.

One of the more intriguing mysteries is the discovery that the earth's magnetic field reversed about 800,000 years ago and has reversed many times previously, a geologist friend of mine, Dr. Robert Nason, concludes. "It only proves that things are funny. No one knows why." In fact, though the magnetic field is weak to begin with, about the same strength as between the opposite poles on a toy horseshoe magnet, scientists are today suggesting that the magnetic field began to wane 10 percent to 15 percent starting as recently as 150 years ago and that it would take about 5,000 years for the magnetic poles to actually reverse. When they do, the insects, birds, and fish—like bees, geese, and whales—that depend on magnetic direction for their annual pilgrimages will be very confused. In addition, there will be other consequences to our earth's system. Our magnetosphere, which extends the power of the magnetic field out in our sky for thirty-seven thousand miles, will protect us less from solar winds and bursts of radiation if the power is reduced, as it now seems to be. Scientists think the magnetic field has to do with the interrelationship of a solid inner core that has the density of iron, with the more fluid outer core of the planet, but, like the reversals, we will never truly know why.

Yet every scientist acknowledges that what we know about any given scientific discipline is dwarfed by what we don't know and that is the great adventure of science. We still do not know, for instance, how human life was created on earth. We do know that we can identify human life by the existence of DNA, but how life developed continues to be uncertain. No less an authority than the codiscoverer of DNA, Francis Crick, has suggested that life could have been sent to earth in meteorites by extraterrestrials. The definitions of science and scientists are as diverse as are the scientists themselves.

Most seismologists in university earth science departments do not want to be known as geologists, and, unlike geologists, do not rely on field experience to theorize about the movement of the earth. They rely, rather, on the computational modeling of previous and projected earth movements, volcanoes, tsunamis, and rates of primary and secondary waves in earthquakes as well as on calculations made in physics, geophysics, math, and electrical, magnetic, radioactive, and gravitational engineering. It is an impressive intellectual discipline. But sometimes these scientists have to make the best of what they do not know, which in this case is an understanding of the *patterns* of earth and water movements that have occurred in geologic time, if indeed these patterns exist. The word *recently* in geologic time could mean more than a hundred million years ago, and a pattern could be something that might be obvious if only we could see clearly the events of three billion and four billion years ago. So every rock that is found is dated and consigned to some period of time that conveys with it some event, and every bit of information is catalogued in a computer for analysis, and if it is seen that the same sort of rock can be found in two different time periods it can be concluded that similar things happened in those periods to create such rocks. If such an observation occurs often enough it can be called a pattern and can be fundamentally useful in predicting the future.

When so many aspects of the origin, movement, and future of our world remain subject to conjecture, there is always room for novel ideas. For instance, it was recently suggested by two eminent Canadian scientists that it is not the movements of the plates of the earth that cause earthquakes but the floating upward of huge plumes of hot and molten rock from the liquid core of the earth (as measured by variations in the planet's rotation rate and gravity field), which cause various types of geological events: volcanoes, ocean disturbances, and movements of the earth's crust.

For example, major earthquakes have occurred in areas where they would not have been predicted based on standard models. Scientists think that the largest estimated earthquake in the United States in recent history occurred in the area of the Mississippi Valley in 1811, with serious aftershocks following in 1812. Boats were thrown over in the river, and many people drowned; entire islands simply disappeared. This earthquake was so large and destructive that Congress passed the first congressional relief act in 1815 to support the farmers whose previously fertile and farmable land had been turned to swamp, sand, and mud. An area of about 1,550,000 square miles was affected, compared with about 93,000 square miles in the 1906 event, and tremors were felt as far away as New York. It is not known how many fatalities there were, but in 1811 the valley was sparsely settled with less than ten thousand people, most living in log houses, which would have sustained the shaking well.

The seismological activity that caused these quakes has never been explained in acceptable terms. There are no plates beneath the area, and there is no fault zone. Seismologists refer to this event as the New Madrid earthquake, having occurred in the New Madrid seismic zone, but no one knows for certain why it should even be considered a seismic zone. Scientists speculate that the earth tried to separate here six hundred million years ago, but failing, a weakness of some kind was created beneath the surface. The USGS refers to this area as a plate boundary zone, which simply means that they are uncertain of any plate boundaries in the vicinity. But recently evidence has been discovered of many substantive earthquakes in a wide area of the southern Midwest from St. Louis to Memphis, an area where more than ten million people live today.

Can such an event happen again? Sure. Will it happen again? No one knows. As Dr. Nason says, "Nature is never overdue. If we think it is overdue it has been modeled wrong."

In any case, the observation and reporting periods cover far too brief a period of time to enable us to discern what might be seen as a pattern. We do know that there continue to be small and medium tremors in the Mississippi Valley and that a quake would kill a large amount of people if it struck at the 1812 level of magnitude.

With the Sumatra earthquake of December 26, 2004 and the resulting tsunami, the world swiftly came to realize the devastating and tragic consequence of great water movement caused by the shifting of plates far beneath the sea. That more than a quarter million died as a consequence will one day become a matter of clinical data, much like the Tangshan earthquake of 1976 that killed upward of half a million in China. These terrible events take place with unpredictable regularity. Scientists can explain only so much before they are forced to simply acknowledge that they will happen again, sometime, somewhere. The most difficult thing to accept in studying great earth and water movements is that there is no comfort to be found anywhere. They cannot be controlled, and very often, particularly if they happen in the middle of the night as happened in Tangshan, they cannot be avoided, outrun, or circumnavigated.

In 1958, a mild earthquake in Lituya Bay, Alaska, caused a landslide into the bay that created a tsunami 450 feet high. Some geologists might call this a *seich,* a standing wave in a closed body of water such as a lake or bay, something like the sloshing of water in a bathtub, a wave that would have swept over buildings forty stories high. Except for the killing of two people in a boat, it was fortunate that the wave headed inland into an uninhabited wilderness area and did not travel out of the bay.

The possibility of great landmasses falling into the ocean is always with us, and huge iceberg droppings, called calvings, occur often without creating great tsunamis. But just recently scientists have found vertical fault lines through the west side of a volcano on La Palma, one of the smaller and westernmost islands of the Canaries. The volcano has a crater that is about five miles wide and a half mile high, which has guided the eruptions that occur about every two hundred years. The last eruption was in 1948, but the newly discovered fault lines have convinced scientists that with some future eruption part of this huge crater will break apart and slide into the ocean, bringing more than a half trillion tons of rock with it. Since tsunamis are created in proportion to the amount of land that falls into the water, this would create a wave mass that some scientists say could be forty feet high if it slides down into the water. Others say that if it crashed into the sea, the wave would be so high it would be unknown to written history, even bigger than the wave at Lituya Bay. It would diminish a little as it crossed the Atlantic, but it would be high enough to do great damage to the high-rise buildings of Boston, New York, and Miami, the cities, among others, it would hit.

Scientists do not know if it will take one, four, or ten eruptions to separate the landmass, only that the separation is inevitable, and there is no known plan to begin to reduce the size of the mountain. The only good news is that volcanoes usually send signals before they erupt, and it would take eight hours for the wave to travel from Africa to the eastern U.S. shoreline. It is not sufficient time, however, to move all the people who would be in its path, and in today's world, most homeland security experts agree that our national evacuation plans are inadequate.

The city of Lisbon, Portugal, was rendered to ruins at 9:40 A.M. on All Saints' Day in 1755, when almost the entire population was gathered in churches. The shaking lasted for ten minutes and is estimated to have been a 9.0 earthquake. A fire started that spread quickly. Many people tried to escape by boarding boats at the harbor. The ocean pulled back from the city thirty minutes later, exposing the sea bottom for a long distance, and then it rushed back to the city in a record fifty-foot wave. The wave also spread across the sea to North Africa and hit Morocco from Tangier to Agadir, killing thousands. In all sixty thousand were killed in Lisbon and ten thousand in other locations.

While we all understand the dangers of living in an earthquake-prone area, what area in the world can we reliably say is earthquake free? Surely the people in the Mississippi Valley feel they are safe, as do the people in New York City. Yet New York City does have a fault line running across 125th Street, of which the great majority of the city's population is doubtless unaware.

People in America have always lived in some of the country's most dangerous places—on the flat tornado fields of the Midwest, on the hurricane-battered coasts of Florida, and on the floodplains of the South. Like many of those in India, Indonesia, and Thailand their choices are often limited by poverty. Being an optimistic populace, however, Americans often convince themselves that lightning never strikes twice in the same place. But, of course, there are many examples where it has. Terrible events in the future are inevitable, and most people regard them hopefully. Yet hoping is not sufficient. The responsibility of all citizens must include an active awareness of the geophysical cataclysms that are possible and ensure that our political leadership is equally aware of and prepared for them. For first responders there is no second chance in a physical crisis. First responders are as well trained in America as in any country in the world. They are highly motivated and committed to duty, and they will always be on duty to protect and to help. But what should be expected from our elected officials whose emergency education is almost always meager? Should we take it for granted that events of the recent past will simply repeat themselves? Surely not. Every crisis in our emergency history has been handled adequately so far—Chicago, Baltimore, San Francisco, Mount St. Helens. But is this the only past that will provide lessons for our future or is there another that we don't know enough about and, consequently, do not even consider? Perhaps that past lies hidden—hidden in the rocks of our geologic history.

CHAPTER 26

Freeman and Pond were breakfasting when the second shock came at around 8:15 A.M. By now they were aware of the great urgency of the situation, for an unsigned telegram had arrived at Mare Island stating that San Francisco had suffered terrible damage. Men had begun to return from liberty as well as the sailors who ran the base's daily market boat, telling of the fires in the Bay City. Stories had also begun circulating about the destruction of and fires in Santa Rosa, a picturesque city fifty miles north of San Francisco and thirty miles inland, in the heart of the burgeoning Sonoma County wine country.

The second shock was not as strong as the first but was still of sufficient intensity to make Lieutenant Freeman realize that this would be a long and memorable day, for the world seemed to be falling apart.

Freeman and Pond felt the shaking of the aftershock—more than ten seconds in length. This second quake would have measured at least a 6 on the Richter Scale, a force sufficient to make them seek a transom between hatchways, a place safe from falling objects. When they recovered their footing, they immediately made a round of their ship. Finding no damage they returned to their breakfast. It was to be their last idle moment for the next three days for soon the base commander, Admiral Bowman H. McCalla, sent for Lieutenant Freeman.

McCalla was a role model for all the sailors of Mare Island, having distinguished himself in the Spanish-American War and, more recently, in the Boxer Rebellion, in consequence of which he was awarded the Congressional Medal of Honor, the Order of the Red Eagle from the kaiser, and a medal from Edward VII of England, for rescuing foreign nationals under siege in Beijing. He was also socially prominent in the California military hierarchy because he had developed a close friendship with General Arthur MacArthur, the former commander of the Pacific Division. MacArthur was something of a hero to McCalla as he had proven himself in the latter part of the Civil War when he had commanded a regiment at the mere age of twenty and won a medal of honor for bravery in the taking of Missionary Hill. McCalla's daughter, Mary, had also just married his son, Arthur MacArthur, Jr. (another son would become one of the greatest military figures in American history, General of the Army Douglas MacArthur), and at the time of the earthquake she was pregnant with a son who would be named Bowman McCalla MacArthur, a name surely that would gain the young man immediate entry into any soldiers and sailors club.

The McCallas and the MacArthurs were given the opportunity to fully enjoy the perquisites of a peaceful military life with the assignment to the Presidio Army Base and the Mare Island Naval Yard, which were among the most coveted postings for servicemen of the time. It was here that the scenic topography, moderate climate, and quietude of a naturalist's haven joined with the cultural excitement of San Francisco to provide as ideal a military life as was possible.

Admiral McCalla, like the army's current Pacific Division commander, Major General Adolphus Greely, was just two months from his planned retirement to Santa Barbara, where he had recently bought a large and beautiful house. But even through his last days he continued to be known as a stern disciplinarian, a skipper who demanded punctuality, efficiency, and absolute adherence to the rules of dress and comportment. He never hesitated to reprimand or court-martial anyone who violated regulations in any way, whether a sailor, enlisted man, or officer. Still, it was the consistency of his fairness that inspired his men's allegiance to his leadership. And though Major General Greely was on temporary leave to attend the Chicago wedding of his daughter, leaving his second in command, Brigadier Frederick Funston, in acting charge, McCalla was very much in control of his navy on this day.

Lieutenant Freeman was eager to take whatever role the admiral would assign him in the earthquake relief effort. The admiral had already sent Lieutenant Commander Lopez into the city on the tugboat *Sotoyomo* to inform the officials of the stricken city that the navy would do all it could to offer aid, and promised that assistance would be forthcoming.

Since the *Perry*'s engines were disassembled, McCalla ordered Freeman to take command of the *Preble,* to gather every doctor, nurse, and orderly on the base along with Past Assistant-Surgeon C. G. Smith and to transport the assembled medical party to the city. There, he was to offer the medical services of the navy to the injured and dying of San Francisco.

With these instructions Lieutenant Freeman became a whirlwind of activity and crisis management. At about 8:30 A.M. he sent Chief Boatswain Daniel Moriarty on the navy fire tug *Leslie,* which was used as a fireboat at Mare Island and the most powerful of the tugs in the harbor, to assist however he could with the firefighting efforts. At 8:45 A.M. Midshipman Pond was dispatched with a detachment of marines, in command of the tug *Active,* with orders to provide whatever relief and order that might be needed. He was also told to tow the *Leslie* if she was overtaken on the water.

Recognizing that there was a shortage of trained warrant seamen, Freeman took on a small crew of six men each from the submarines SS *Grampus* and SS *Pike*—tiny Holland-class subs that were said to be more dangerous to their passengers than any enemy ship. A third sister ship, the torpedo boat destroyer *Paul Jones,* had a full crew but had been put into service patrolling the harbor to prevent any ships from fleeing the jeopardized city with critical cargo—foodstuffs, clothing, tools, and mechanical instruments—that might be used to save lives.

At around 9:00 A.M., Freeman finally shoved off for San Francisco. The trip was twenty-four miles as the crow flies but thirty-five miles by sea, and even in the fastest boat in the navy it would take seventy or so minutes. In the race to the city, the faster tug *Active* passed the *Leslie,* and at about 9:30 A.M. Midshipman Pond threw a line to the slower craft and towed her.

Even from a great distance Freeman could see the darkened skies at first, and with each knot closer to shore he realized that the situation was growing ever more ominous. He began thinking of other catastrophic fires in history, of London and Moscow and Chicago, and most recently in Baltimore, and was thankful that McCalla had had the wisdom earlier that morning to send in as much gunpowder as they could find on Mare

Island to use to create firebreaks. Like many in San Francisco that day, Freeman thought the idea of using dynamite was a viable one.

The skipper held binoculars to his eyes when they reached a few miles out and scanned the horizon with them for most of the remaining trip. In studying the mass of flames he could see that the fire was developing winds and currents of its own. Lieutenant Freeman saw immediately that the entire city could burn and that his job was not only to help out at whatever particular sites he could, but also to try to save San Francisco from destruction. But how? He knew from his first aid courses that in treating potentially mortal wounds, maintaining the blood supply to the heart was the primary consideration, and the same was true now—he had to protect the main artery of the city, its seaport and railroad yards, the heart's blood of a vital commerce. The city would never be able to rehabilitate or survive the future without them.

Once again the prescient fire chief, Dennis Sullivan, had pleaded with the board of supervisors in his annual report of 1903–4 to heed this very matter. "I again call attention," he wrote, "to the most urgent need of better fire protection for the valuable water-front property and the shipping interests, and in order that this might be accomplished a light-draught, high-power fire boat of good speed and large pumping capacity, should be provided."

The *Preble* overtook the *Active* and the *Leslie* just before they reached the docks, and Freeman noticed that the seamen were stunned by what they saw as they approached the city before them. The city of shore leave delights, the wide-open city that gave rise to memories of wild and raucous times, the famous city, was now a montage of ruination and terrified humanity. A huge black cloud, almost the size of San Francisco itself, was billowing upward, framing the wildly dancing oranges and yellows of the flames below. No one had ever seen anything like it, and it seemed the fire had already grown larger than those of Chicago, of Baltimore, or of the great fires of ancient cities. Midshipman Pond rushed to his duffel bag and rummaged through it until he found the Brownie camera he always carried to take the travel photos he would send home to his mother and father, wherever in the world an admiral's travel might take them. He wanted a record of what he was seeing because he did not think anyone would ever believe from his description alone that the morning sky could be so utterly black.

The *Preble* docked at a wharf just east of the Ferry Building and tied up next to the *Sotoyomo*. Lieutenant Freeman ordered the raising of the hospital flag and then reported to Lieutenant Commander Lopez. Freeman was ordered to stand by until Lopez returned from a surveillance trip to Goat Island (now Yerba Buena Island), where there was a military infirmary.

As he waited Freeman sent a messenger to find a fire department official to ask where the two firefighting tugs, *Active* and *Leslie*, were most needed. In the meantime, Dr. Smith and his medical party disembarked to enter the city to determine where their medical assistance would be most useful, but they could find no city officials anywhere. And because of the great heat and the flames, they could not set out in the direction of the civic center.

A battalion chief from the San Francisco Fire Department directed the sailors to Pier 8 at the foot of Howard Street, where an out-of-control fire was being fought by the firefighters. Seeing that the area of the fire was growing, Freeman then decided not to wait, as ordered by Lopez, but to proceed to where he and his men could do the most good. Lopez would understand his decision, Freeman concluded, and under the circumstances there would be no consequence in countermanding a senior officer's instruction. It was like Freeman to heed his own call to action.

At Pier 8, Dr. Smith and his medical crew set up treatment tables and chairs in the smoky and cinder-filled open air. By now hundreds of people, most in torn and disheveled clothing, had made for the shoreline, where they believed they would be safe from the fire. Many of them were suffering from serious injury, and Dr. Smith made arrangements to transport them to the Goat Island infirmary.

Lieutenant Freeman directed his men to the hose lines that were coming off the *Active* and the *Leslie*, now tied next to the *Preble*. There would be good, strong saltwater streams coming from the pumps of the tugs, which had been specially converted by the navy to fight fires. He could see there was a group of firemen just off the docks at Steuart and East streets, near the Ferry Building two blocks away. Two of the assisting companies from Oakland were drafting from the bay and spraying the Ferry Building to keep it cool and safe. Nearby fires were raging, and their heat was coming at the firemen and the seamen in waves. Freeman noticed a battalion chief and ran to him, explaining that the seamen were going to join forces with the firemen.

"Just hit the exposures," the chief replied. "We'll try to stop it here."

Lieutenant Freeman nodded and turned to his group. "Heave to, men!"

he shouted, lifting a section of limp hose and directing the seamen forward, "and, sock it to 'em."

With his forces in place, Freeman looked to his rearguard and shouted, "Start the water." The pumps began, and with the water came the start of Lieutenant Freeman's effort to save the city.

CHAPTER 27

As the world-famous tenor Enrico Caruso jumped from his bed in room 580 of the Palace Hotel, the walls surrounding him began to vibrate wildly. His valet was in the adjoining room, along with the singer's thirty-eight traveling trunks, the revolver he had bought for protection in "the Wild West," and a photo of himself and Teddy Roosevelt, which he had been given just a few weeks before by the president himself. Caruso dusted himself off, took the photo of him and the president in case he needed an identifying passport, and went to the street. He walked four blocks to the St. Francis Hotel on Union Square, where huge crowds were already gathering with their possessions on the open expanse of grass. Caruso went to the hotel's dining room, ordered bacon and eggs, and, it is reported, tipped the cook $2.50. It would be a difficult day and night for him as he attempted to make arrangements to flee the city, all the while commenting on the double tragedies of the earthquake and the eruption just eleven days earlier of Vesuvius, which had killed more than two thousand people near his home in Naples.

CHAPTER 28

When Bridget Conroy felt the earth shaking, she immediately tried to stand. She held a wall and waited for the quaking to stop. It *would* stop, she knew. She had been living in San Francisco for a quarter of a century, after all, and had been through many of these temblors.

When she felt she could control her footing, she lighted the gaslight and searched for her robe. She then rushed into the children's room and saw that they had slept through it, youngsters safely traveling through their dreams. She looked out the window, where it was growing light. Though less than a half mile from city hall, she did not hear the huge dome tumbling. There was pandemonium throughout the neighborhood, but here on Harrison Street everything seemed quiet and orderly and the earthquake was just another passing memory. Satisfied that everything was safe, she returned to her bed. She did not notice, though, that the gaslight had gone out. The gas had surged and extinguished the flame, but the gas itself continued to flow. It flowed as she closed her eyes and fell again into a deep sleep. She slept easily for the next twenty minutes or so until she was unconscious altogether, and the life of a hard-working immigrant began to fade into a family's memory. In the years that followed many of her blood would serve the people of San Francisco: police officers, firefighters, lawyers, commissioners, and, ironically, the city's current director of the office of emergency management.

They did not discover her body until much later in the morning, as the fire began to encroach her son's home, and people were gathering as much as they could carry for the evacuation.

CHAPTER 29

Photographers abounded in San Francisco in 1906, and for the first time in America a truly historic catastrophe was fully documented by hundreds of photographers, both amateur and professional.

On the third floor of 12 Jones Street, in the shadow of the large Spring Valley water reservoir on the top of Washington Street, H. M. Griffith and his son T.W. were looking out of their window at perhaps the most expansive view in all of San Francisco, a panorama looking north, from Fort Mason to the Ferry Building. They were amateur photographers and the vista never failed to move them, but on this day their perspective was more than panoramic—it was historical. Though renters, the Griffith family was solidly of the commercial class, and their residence on Nob Hill must have been large, at least with room enough for a Chinese servant. Apart from that, hardly anything is known of them, and the Griffiths seem to have faded into the obscurity of the city's past.

It was a clear morning, and the shaking was worrisome, but their house, as did all houses built on the rock formations on the high hills, sustained the earthquake well enough. There might have been some breakage of dishes or a frame might have fallen from a wall. Mr. Griffith surveyed the landscape before him and pointed to the far right, just over Chinatown and the business district, where smoke was building. The two men probably asked each other, as all lovers of photography would have naturally asked: Is there film in the camera?

These amateur photographers, however, did not realize that their photographs would be lost in the walls of a house for decades. They came to light only in 1963, when a San Francisco construction worker reported to a newspaper person that he had found some photographic negatives in the wall of his house, and they seemed to be of the fire. The photos then made their way to the light of day and are today among the thousands of photographic treasures to be found in every museum of San Francisco.

CHAPTER 30

Like every other San Franciscan, Brigadier General Frederick Funston awoke with the shaking. His home, at 1310 Washington Street, seemed to be rolling in waves, and the general immediately knew by the length of time that was passing that the temblor was significant enough to require his full attention. His house stood at the northern base of Nob Hill in one of the finest neighborhoods in San Francisco, and because it was made of wood and was on firm pilings it suffered no damage, at least in the initial earthquake. Funston dressed quickly in civilian clothing and went into the streets to survey the damage, thinking that he would miss his mandatory morning coffee. He took note of the effects of the quake on some of the houses around him and saw the beginnings of several fires. At that moment he began to think in the concentrated, focused terms of military strategy.

Throughout history, cities and even nations have at times become victims of simple bad luck. In 1660, for example, Charles II approached his reign with a deserved sense of optimism. After all, he had survived a five-year rule of religious ruthlessness by Cromwell and two ineffective years of leadership by Cromwell's son, Richard, until he had finally been able to take the throne to bring prosperity once more to England. But he could not have been prepared for the appearance of the Great Plague of 1665, a disaster that was immediately followed by another, the Great Fire of London of 1666. Similarly, San Francisco experienced two strokes of misfortune in the earthquake and fire of 1906. In the passing of its respected fire chief, Dennis Sullivan, and the absence of its area's commanding officer, Major General Greely, power either fell into the hands of or was seized by Funston.

The first recorded fire in San Francisco—then a city of thirty-one frame houses, twenty-two shanties, and twenty-five adobe structures—was a simple brushfire that occurred in January 1847. It caused very little damage but gave rise to the city's first fire regulation—a fine of twenty-five dollars to any man who started a brushfire without permission.

Until that terrible day in 1906 the worst toll of death and injury in San Francisco had been on November 29, 1900, when the roof of the Pacific Glass Works collapsed during the annual Stanford–University of California football game. Eighteen men and boys who had climbed onto the roof of the building to get a free view of the game were killed and more than eighty-three were seriously injured—nearly half of them as they fell onto the top of the huge furnaces that had taken a month to heat to the high temperatures needed to blow glass.

But the earthquake would challenge the department as no other fire department had ever been tested before. The two ranking fire chiefs, John Dougherty and P. H. Shaughnessy, were uncertain as to who would replace Dennis Sullivan as the chief engineer. Because Dougherty, who was in his mid-sixties, had been planning to retire in June, Shaughnessy was the obvious choice, but both men knew that the appointment would be up to Mayor Schmitz, who took no consequential action without the approval of the political boss and head of the Union Labor Party, Abe Ruef. Given that uncertainty, neither man was willing to take absolute control of the emergency, as was the legal right of the fire chief. Chief Dougherty was named the acting fire chief during the conflagration, but he did not assume the level of command that Sullivan most probably would have.

This legal right had been established in 1850 when the city's first fire chief had the city's first mayor escorted from a fire scene after an argument regarding authority. The mayor, John W. Geary, wanted to fight the fire one way while Fire Chief Frederick Kohler wanted to direct the efforts. Geary (for whom the boulevard in San Francisco is named), who had made a personal gift to the city of Union Square, was a formidable opponent—tall, a man who would go on to be a hero at Gettysburg and a two-term governor of Pennsylvania. He wasn't used to losing fights.

But Kohler was an experienced former New York City firefighter who knew that in emergencies there should be just one commander, for the absence of clear management in a crisis of public concern means gambling with the lives of citizens, and particularly with the lives of rescue workers. He had Geary arrested that night for interfering with fire department operations and so set the legality of the fire chief's hegemony at emergencies. (This line of incident command exists today in most American cities, with some notable exceptions like New York and San Francisco, where incident command respectively is shared with the police department or controlled by the mayor.)

In 1857, a fire occurred during which a San Francisco fire chief ordered his men to pull down a small house to create a firebreak, and the owner of the house sued the chief for illegal confiscation of property. But in *Cuneo vs. Gerry* decided that year, the court confirmed the right of the fire chief to destroy buildings in the interest of public safety.

Had Dennis Sullivan survived the earthquake he would undoubtedly have declared himself the absolute manager of the catastrophe and asserted his right of office. He was an engineer, whose area of expertise was hydrolic engineering, and the lack of water was exactly the problem San Francisco was facing. Instead of confident and certain leadership, the city was faced with two competing and reluctant chiefs, a corrupt and boss-manipulated puppet mayor, and an army commander—a battle-scarred hero with no experience of peacetime action anywhere—whose authority became unquestioned.

That person, Brigadier General Frederick Funston, unhesitatingly took control of the crisis in his own hands.

Many differing accounts have been written concerning the actions taken by the U.S. Army in the first hours of the earthquake and fire. Funston himself wrote two reports on the disaster that differ in small but important points—one published in the July *Cosmopolitan* and likely written in May 1906 and the other, his official record of the events, submitted in July to his immediate superior, Major General Adolphus Greely.

In these accounts Funston either excuses actions of his that had been criticized, such as his extreme reliance on the use of dynamite in attempting to control the fire, or he denies them outright, such as the accusation that he had illegally declared martial law. Many of the military reports submitted at the time by various junior officers were not declassified or uncovered until the 1970s and 1980s, and so there were no contemporaneous opportunities for corrections to or rebuttals of Funston's version of the response to the earthquake. Indeed, one of the most important statements of the military role in 1906, that of Lieutenant Freeman, was inadvertently classified for more than seventy years. In addition, other critical eyewitness records, such as an original, unexpurgated monograph written in 1931 by Lieutenant Freeman's second in command, Midshipman John E. Pond, have only recently been discovered, again after seventy years. So, consistent with the old adage that truth will out, the story of the military's role in the great earthquake and fire can now finally be told.

When he returned from the Philippines, Funston had first been given command of the Department of Colorado, and then the Department of Columbia (which consisted of the northwestern states). But by 1906, he had been appointed commander of the prestigious Department of California, reporting directly to the commander of the Department of the Pacific, Major General Greely. With Greely out of the state, he was therefore left in charge of all the army's troops and property in the states of California, Hawaii, Nevada, Oregon, and Utah.

It is important to follow Funston closely during these first few hours of the fire, for it was during this time that he determined, as the senior army officer in the city, to take command of the emergency.

As he walked through the streets of San Francisco in civilian clothes, his initial responses were no different from those of any other citizen, simply wondering what havoc had been wreaked upon the city. There was no transportation, for not only had the quake distorted the tracks of the cable cars, but it had also knocked out the electricity needed to power the underground network of cables that powered them. Nor was there any telephone service with which he might locate a car and a driver. Funston had no choice but to travel on foot, though he did from time to time try to stop passing automobiles.

Seeking a strategic vantage point he walked south on Jones three blocks, then east on California Street another three or four blocks to the Nob Hill summit. It was a peculiar choice because the situation on California Street, just three short blocks from his home, would have been higher, and in 1906 would have offered a better panoramic view. In any event, he wrote of seeing several columns of smoke rising from the region south of Market Street, the hardest hit section of the city, and a few fires beginning to rage in the banking district, near the foot of Market, close to the Ferry Building. He walked downhill on California eight blocks to Sansome Street, where he observed other fires and noted that the fire department was helpless, "owing to water-mains having been shattered by the earthquake."

In point of fact, water was available at this location. Many hydrants were working in the area, and a twenty-five-thousand-gallon operational cistern stood at that very corner. But, most important, the fire department's own accounts and also the report said to be written by Chief Shaughnessy himself, indicate that firefighting operations at California and Sansome did not get under way until much later in the morning.

The general's first thought and determination were to protect the city by guarding the police lines. But it can be argued that a city threatened by conflagration benefits more from a commitment to save lives than property. Funston's rationale was based on a military officer's responsibility to ensure the survival of the city's federal buildings—the post office, the mint, and the customs house among others. The reason he took pains to report his peregrinations in the early morning, during which he supposedly witnessed the many fires, dry hydrants, and helpless firemen, was to justify bringing federal troops into San Francisco without having the civic authority to do so.

In the *Cosmopolitan* article Funston recalled that he now returned home for a cup of coffee and to instruct his wife to pack their belongings to be moved to a military installation. In his report to General Greely, however, he indicated that he ran and walked the eleven blocks to the military stables at Pine and Hyde streets, from where he sent a messenger with a note to Colonel Charles Morris, the commander of the Presidio, three miles distant, and to Captain Meriweather Walker, the commander of Fort Mason, a mile away, directing them to report with their entire commands to the chief of police at the hall of justice at Portsmouth Square. This note has never been found, and, in fact, the soldiers were apparently first headed for city hall; they changed their destination to the hall of justice only after having encountered the mayor on Van Ness. It is certainly possible that General Funston sent the troops directly to city hall, to take "control" if need be, but on reconsideration thought it better to *claim* that they had been directed to the hall of justice to assist with policing duties—a subtle distinction in countering any accusations that he ordered martial law to be in effect.

When, after a search, the messenger who had been sent to the Presidio finally found Colonel Morris, it was reported that the feisty colonel told him to inform General Funston—whom he called *that newspaperman*—that he should "look up his Army regulations, and there he will find that nobody but the President of the United States can order regular troops into any city." Indeed, in the Chicago fire thirty-five years before, the great Civil War general Philip Sheridan ordered his troops into the city, only to have that action vigorously protested by Governor John Palmer of Illinois. In that case, just a few soldiers were left behind to patrol while the rest withdrew.

Ignoring the colonel's response, the messenger himself ordered the bugler to play the mustering call to assemble the troops of the Presidio. With this insubordination the officer placed himself in serious legal jeopardy, but the colonel then acquiesced as he saw his soldiers forming into ranks. He might well have concluded that obeying the order was inevitable, and he marched the soldiers as far as O'Farrell Street, where he came under the command of General Funston. The soldiers had their bayonets fixed and gleaming at the end of their rifles. They were prepared for action, and General Funston approved of their manner of readiness.

At Fort Mason, the messenger had to bang on the door to awaken Captain Walker, who had returned to sleep after the earthquake. Alarmed by the urgency of the messenger's information, he quickly mustered his 5 officers and 150 soldiers, again with their bayonets fixed, and began the trek into the city.

Prior to Funston's summoning of the troops, Chief Dougherty had sent his own messenger to the Presidio to ask Colonel Morris for dynamite—a messenger who had been much more warmly received than General Funston's. In all likelihood, Chief Dougherty had discussed the alternative of dynamiting with the mayor and Funston at an earlier meeting at the hall of justice, before Dougherty sent his messenger to Colonel Morris. Dennis Sullivan and Chief Dougherty had spoken many times about the need for a dynamite-driven alternative to fight fires in the city, although there is no record that anyone in the fire department, with the exception of Sullivan, or the army had ever studied the precise procedures for creating firebreaks with explosives. "This town," Sullivan had said in a speech, "is in an earthquake belt. One of these fine mornings we will get a shake that will put this little water system out, and then we'll have a fire. What will we do then? Why, we'll have to fight her with dynamite."

After speaking with the messenger, Morris directed Captain LeVert Coleman to acceed to his request, and Coleman immediately rounded up forty-eight barrels of gunpowder—not dynamite—that were put on a caisson and delivered by Lieutenant Raymond Briggs to Mayor Schmitz, who had recently arrived at the hall of justice.

Lieutenant Coleman then loaded another two wagons with gunpowder, this time with three hundred pounds of dynamite, which were actually borrowed from a contractor doing work at the Presidio. In total, then, about 95 percent of the explosives sent by the army to be used for firebreaks consisted of gun cotton, black powder, and a substance called dynamite powder, all three of which were classified as low explosive, as opposed to the high explosive of dynamite. Hardly anyone in San Francisco understood the nature of these materials, or that low explosives had an undeniable tendency to start fires.

When Captain Coleman delivered the explosives to Mayor Schmitz, he noted that the superintendent of the California Powder Works, John Bermingham, was standing next to the mayor. General Funston was also present. Bermingham had delivered some additional dynamite and, being an authority on it, had volunteered to supervise its use. Coleman also noted that Bermingham was drunk, an observation corroborated in other accounts made during that and the next day. One report states unequivocally that Bermingham's use of dynamite had caused the loss of life.

The fire along Montgomery Street (now Columbus) had been spreading because of the discovery of empty or sand-filled cisterns. The firemen faced disappointment after disappointment as they were forced south and west, for whatever breezes that flew over the bay water pushed them back and pushed the fire across Clay Street.

The mayor, with General Funston at his side, sent Lieutenant Briggs and a fire department battalion chief, Michael Murphy, to Clay Street to a location between Sansome and Montgomery, where Chinatown and the financial district meet, to begin creating firebreaks. There they encountered another battalion chief, Michael O'Brien, who ordered three men from Engine 28 to assist them: Lieutenant John McGowan, August Stoffer, and a hoseman, Albert Bernstein.

John Bermingham was evidently the only expert explosives professional on the scene, but his drunkenness rendered him useless if not dangerous. LeVert Coleman was the Fort Mason ordnance officer, but neither he nor Briggs was a dynamite expert, and he had no blasters in his command. Yet in his official report he states, "General Funston and the Mayor . . . placed me in charge of the work of handling all the explosives."

None of the firemen, it seems, had any real expertise in dynamiting, either, but the determined team bravely set to work, first on the south side of Clay and then at the areas at Commercial and Sacramento streets. In one sense, it was on-the-job training, for over the course of the next two days they would develop a keen understanding of the proper methods of bringing buildings down with dynamite.

The team was given a strong and clearly defined command by Funston and the mayor. Because the leadership was concerned about being accused of unnecessarily blowing up any buildings, and also because of the general uncertainty of what it was they were undertaking, they ordered that only those structures which had already come in direct contact with the fire be destroyed.

This was a decision of major consequence, and if ordering the dynamiting of buildings by people who were not expert was Funston's second mistake, this was his third.

Had Dennis Sullivan been present to supervise the dynamiting, he would have known that it is the building several properties away from the fire on the leeward side that should be brought down because its destruction would create a significant firebreak to stop the fire when the wind pushed it in that direction. Furthermore, there would be sufficient time for the firebreak area to be cleared of as much debris and flammable constructions, like window casings, beams, and even furniture, as possible for the general rule of cleanliness in fire prevention would apply equally to the area of firebreaks.

Chief Shaughnessy, in his official statement in the municipal report, was to write:

> Our department cannot be held responsible for the vast destruction of property occasioned by that conflagration. Our men battled with the flames until exhausted and it is due to their heroic and splendid efforts that so great a portion of our beautiful city was saved from destruction.

This was his way of saying that the fire department had had no direct power over the decisions that were made in fighting the fire and managing the catastrophe, for he knew as well as any man that the fires had burned slowly and that it was the dynamiting that created much of the destruction.

As the dynamited buildings on Clay, Montgomery, and Sacramento fell, not only did they quickly catch fire, but some of the explosives sent burning embers flying across the street, spreading the fire in the very manner the explosions had been intended to prevent. The dynamiting, therefore, not only fueled, but worse, was actually incendiary, giving rise to an entirely new fire on Sansome Street, one that would eventually take all of Chinatown.

As Coleman said in his report, "... although we were able to check the fire at certain points, it outflanked my party time and again, and all our work had to be begun over." In other words, the firefighters were forced to deal with an almost hopeless situation. At this time they were facing a fire just north of Market Street that was extending to join with the fires just south of Market and together would form a wall of fire a half mile long.

Just as the fire began to cross Market Abe Ruef suddenly appeared at Montgomery Street. He owned the Commercial Hotel there, an eight-story structure that he recently had built, and he suggested to all in the uniformed forces, military or municipal, that it would be in no one's interest to have it dynamited.

CHAPTER 31

When the quake struck that morning, the neighborhood around the residence of San Francisco's mayor, Eugene Schmitz, suffered almost no damage. In fact, the mayor thought it was an insignificant event, and, like the widow Conroy and Captain Walker, went back to bed, having no idea of the destruction that had been wrought in the South of Market area or that his beautiful city hall lay in ruins. In fact, he was only informed of the true extent of the damage when two men, John T. Williams and Myrtile Cerf, showed up on his doorstep at 6:00 A.M. with the news, forty-five minutes after the earthquake. They had already gone by car to city hall expecting to find Schmitz there, but not finding him went to the mayor's residence. Schmitz was stunned by the news of the scale of the disaster and immediately drove to see the devastated city hall.

On the way, according to Schmitz—an account corroborated by both Williams and Cerf—when they reached Van Ness Avenue to head south, they encountered Fort Mason's commander, Captain Walker, and a little more than a hundred troops. Schmitz asked Walker where he was taking his men.

The officer told him that they had been ordered to march to city hall. Funston, of course, claimed in his account that he had ordered the men to the hall of justice, nearly twenty blocks away.

Schmitz then explained to Walker that city hall was in ruins, "and directed him to report to Chief of Police Jeremiah Dinan at the hall of justice." The mayor no doubt saw the armed troops as playing a law enforcement role and thought that Police Chief Dinan would be best suited to handle their deployment.

CHAPTER 32

It is uncertain when Mayor Schmitz first encountered General Funston that morning, but their earliest recorded meeting was at about 7:00 A.M., when Captain LeVert Coleman, who had been placed in charge of dynamiting buildings, reported to the hall of justice, where he found both men. The general, Coleman wrote, "placed me in charge of the work of handling all the explosives." His notation implies a shared command at the hall of justice, which raises the question whether, previous to this encounter—that is, at some point earlier in the morning—General Funston had informed the mayor that he had already ordered in the troops and, that being a fait accompli, he now needed the mayor to publicly suggest that the city needed the army's help. The crisis, Funston must have realized, had to be acute and life threatening—it had to be more than just the fire that threatened, in other words—if he was ever to be able to rationalize the summoning of armed federal troops into an American city and to answer the criticism that was certain to be leveled against him.

Mayor Schmitz, meanwhile, had just sent off a message to Governor Pardee to ask him to prepare a response for any exigency, and one to Mayor Mott of Oakland asking for assistance from that city's fire department. Because the residents of Oakland were not such late-night people or such early risers as San Franciscans, there were not many cooking fires going when the earthquake struck, and only seven alarms for fire were sounded there after the quake. After handling their own fires easily, the Oakland Fire Department sent help to its sister city across the bay.

Mayor Schmitz also directed that the soldiers and police close down every saloon in the city—no small order to fill for there were about a thousand of them, more than the number of grocery stores. Certainly, San Francisco had a reputation as a town of constant celebration, and it also had its cast of bums and drunks but, in fact, most of the saloons were centered within the already burning Embarcadero and harbor districts. In retrospect the prioritizing of liquor control seems another example of the misuse of manpower in an extraordinary crisis. It is evident that the mayor's later order for the actual destruction of liquor—more than six hundred thousand dollars' worth of private property, it was later ascertained—was a response to a request made by Colonel Morris to General Funston. (That Colonel Morris was the first to suggest this action was related in General MacArthur's investigation and report made later that year.)

Finally, the mayor issued an unprecedented command to the police chief: "As it has come to my note that thieves are taking advantage of the present deplorable conditions and are plying their nefarious vocations among the ruins in our city, all peace officers are ordered to instantly kill anyone caught looting or committing any other serious crimes."

The mayor's proclamation was tantamount to declaring martial law, and martial law is always accompanied by a military presence. In the hundreds of accounts of the catastrophe available, given by members of the police and fire departments as well as by the general population, martial law was assumed to have been in place for the ensuing four days. In fact, both the general and the mayor were entirely without legal right or justification to use such force.

Of the early morning, Funston writes, "I realized then that a great conflagration was inevitable, and that the city police force would not be able to maintain the fire-lines and protect public and private property over the great area affected. It was at once determined to order out all available troops not only for the purpose of guarding federal buildings, but to aid the police- and fire-departments of the city."

It is worth noting his phrasing here: "It was at once determined to order out all available troops. . . ." In his report to General Greely and in his article for *Cosmopolitan,* Funston frequently uses this personal distancing afforded by the passive voice.

General Funston was well aware of the fact that the only person authorized to order federal troops into an American city is the president of the United States and, then, historically, only after consultation with the relevant state's governor. He must likewise have been familiar with Clause 2, Section 9 of the Constitution of the United States, which reads: "The privilege of the writ of habeas corpus shall not be suspended, unless when in cases of rebellion or invasion the public safety may require it."

Even during the great Civil War riot in New York in 1863, when the national guard was turned out, the requests for federal troops and the imposition of martial law had been considered and rejected. Martial law had been declared just three times in our nation's history prior to the earthquake: in New Orleans in the War of 1812, during the Civil War, and in 1892, when striking miners rioted and blew up a mill, killing a man, during a strike at Coeur d'Alene, Idaho. Since then it has been invoked only three other times: in 1914, when President Wilson sent in troops to assist the national guard in the coalfield wars of Colorado; in 1934, when Governor Frank F. Merriam of California requested troops to quell rioting dockworkers during a San Francisco strike; and in the protectorate of Hawaii during the period after the Pearl Harbor attack in 1941.

Undoubtedly, the most famous court case involving martial law in American history also involved an illegal use of it. The case had to do with the appeal of a Confederate sympathizer named Milligan who had been sentenced to death in Indiana under Abraham Lincoln's imposition of martial law during the Civil War. The Supreme Court declared that Lincoln had overstepped his bounds for, the court reasoned, "Martial Law . . . destroys every guarantee of the Constitution." It also stated that "the officer executing Martial Law is at the same time supreme legislator, supreme judge, and supreme executive," and issued to the president a resounding warning. "Civil liberty and this kind of martial law cannot endure together; the antagonism is irreconcilable; and, in the conflict [the Civil War], one or the other must perish."

The court did acknowledge, however, that the president could declare martial law when the circumstances warranted it, when civil authority could not function: "If, in foreign invasion or civil war, the courts are actually closed, and it is impossible to administer criminal justice according to law, then, on the theatre of active military operations, where war really prevails, there is a necessity to furnish a substitute for the civil authority . . . to preserve the safety of the army and society."

Given these parameters and considering the availability of 726 uniformed San Francisco police officers in San Francisco on April 18, 1906, there was no demonstrable necessity for the imposition of martial law. Funston described the conditions that early morning: "The streets were filled with people with anxious faces, all turned toward the dozen or more columns of thick black stroke rising from the densely populated region south of Market Street. The thing that at this time made the greatest impression on me was the strange and unearthly silence. There was no talking, no apparent excitement among the near-by spectators; while from the great city lying at our feet there came not a single sound, no shrieking of whistles, no clanging of bells. The terrific roar of the conflagration, the crash of falling walls, and the dynamite explosions that were to make the next three days hideous, had not yet begun."

That General Funston was thought of as a great military leader there is no doubt, and it can justifiably be argued that he saw an opportunity on April 18 to exercise that leadership, recognizing that this was a crisis that would receive international attention. On this day, General Funston thrust into the City of San Francisco, a city at peace and in the midst of a significant life-threatening emergency, 1,700 troops (the 600 troops from the Presidio and Fort Mason were supplemented as the day wore on by another 1,100 troops from other military camps) ready for battle as if every encounter of the day would be with an enemy. But even though the notion of "law and order" implies law first and order second, it was a time in American history when people placed a premium on order, particularly in the West, where relatively little time had passed since quick-drawing sheriffs had been hired to clean up the lawless towns long before anyone ever thought about bringing in a judge. Had Funston sent 1,700 troops into the city to assist in the firefighting effort, its chance of surviving the fire with relatively little damage would have been greatly increased. But all the leaders of the city—the mayor, the eighteen members of the city's board of supervisors, and the members of the special group the mayor appointed to help in the crisis called the Committee of Fifty—seemed to support the general in his view that the priority was to protect the property of the good citizens of San Francisco from bands of marauding thieves and murderers. There is no evidence that anyone objected to this strategy, at least not in the first two days.

In an ironic footnote, General Greely, in his May 17th official report of the emergency, having had ample time to study the event, refuted Mayor Schmitz's and General Funston's rationale for bringing in the troops. "The terrible days of the earthquake and fire in San Francisco," he wrote, "were neither accompanied nor followed by rioting, disorder, drunkenness (save in a very few cases) nor by crime. The orderly and law-abiding conduct of the people rendered the maintenance of order a comparatively easy task."

Jack London, who also wrote an account of the fire, made similar observations about the demeanor of San Francisco's citizens. "As remarkable as it may seem," he wrote, "Wednesday night, while the whole city crashed and roared into ruin, was *a quiet* night. There were no crowds. There was no shouting and yelling. There was no hysteria, no disorder. I passed Wednesday night in the path of the advancing flames, and in all those terrible hours, I saw not one woman who wept, not one man who was excited, not one person who was in the slightest degree panic-stricken. . . . Never in San Francisco's history, were her people so kind and courteous as on this night of terror."

CHAPTER 33

The propriety of the mayor's shoot-to-kill order could have been questioned by others who were present that morning, who might have counseled the mayor against taking so drastic a measure. Among them was Franklin Lane, a city attorney, and Judge John Hunt. But the mayor and his friend Garret McEnerney, a lawyer, went ahead and drafted the public warning, which was the first information promulgated by the Schmitz administration during the quake.

CHAPTER 34

Rudolph Spreckels took his carriage one block from his home down to Van Ness, and headed toward city hall, sixteen blocks away, making his way through the hordes running through the streets. He stopped at his father's home at Sacramento, one of the largest of the San Francisco mansions. Like most of the fabulous homes in the city its architecture was a hybrid, half wedding-cake Victorian and half German baronial, with battlements, a mansard roof, widows' walks, and grand entry steps.

PROCLAMATION
BY THE MAYOR

The Federal Troops, the members of the Regular Police Force and all Special Police Officers have been authorized by me to KILL any and all persons found engaged in Looting or in the Commission of Any Other Crime.

I have directed all the Gas and Electric Lighting Co.'s not to turn on Gas or Electricity until I order them to do so. You may therefore expect the city to remain in darkness for an indefinite time.

I request all citizens to remain at home from darkness until daylight every night until order is restored.

I WARN all Citizens of the danger of fire from Damaged or Destroyed Chimneys, Broken or Leaking Gas Pipes or Fixtures, or any like cause.

E. E. SCHMITZ, Mayor

Dated, April 18, 1906.

ALTVATER PRINT, MISSION AND 22D STS.

The imposing structure was set into a deep moat surrounded by a two-foot wall surmounted by eighteen inches of ornate iron fencing.

Like all men who have created great fortunes in America, Rudolph's father, Claus Spreckels, was single-minded, competitive, and determined to grow each of his many successes into something bigger. Born in Darmstadt, in the independent kingdom of Hanover (now part of Germany) in 1828, he came to America to escape a military conscription when he was just twenty. He wound up in Charleston, South Carolina, where he worked in a grocery, buying the store from a willing boss in just eighteen months. He met and married Anna Mangels before moving to New York, where he began a wholesale grocery business, which had a moderate success until 1856.

Reading of the rapid growth of the West since the Gold Rush, Claus went to San Francisco, and in a short time he built and was running a thriving grocery business on Pine Street, just a few blocks north from Rudolph's office. After a year's success, Claus recognized that there was more money to be made among the heavy-drinking miners and cowboys by selling beer, and he created the Albany Brewing Company. Claus said about himself, "I never have gone into anything unless I could have it all my own way." The brewing business produced prodigious profits, and seeing how much he was paying for sugar in the West, Claus resolved to open a sugar refinery, buying the raw product from Hawaii. He had just moderate success with the Bay City Refining Company and realized he did not know as much as he should have about growing and refining sugar. So he returned to Germany to better learn the business, particularly the secrets of refining sugar beets. This education was to create the foundation for one of the great fortunes of American history, and in a few short years he was known as the sugar king.

Claus and Anna eventually had twelve children, but seven died at birth or in childhood. Family folklore has these children succumbing to the flu epidemic of the 1860s, and some of them probably did. The eldest, John D., born in 1853, would be a partner with his father in most business ventures: He was the publisher of the *San Francisco Call* until the family sold it to the Hearsts in 1912; was the founder of the Oceanic Steamship Lines, which had contracted routes to Hawaii, Fiji, Samoa, and Australia; and owned much sugarcane property, including fully one third of the island of Maui.

The next son, Adolph, born in 1857, became a famous boulevardier in San Francisco, finally at fifty-one marrying the artist's model Alma de Bretteville, whose thinly draped figure sculpted by Robert Aitken dances on top of the Dewey Monument in Union Square. She alone did not bring Adolph happiness, for he had multitudinous business interests, including the Monarch Oil Company, which was bought by Standard Oil, and a baron's chest of property deeds. With his brother John he played a significant role in the development of San Diego, where they owned the Hotel del Coronado, much commercial property, the streetcar companies, wharf rights, shipping rights, and two railroads.

The next brother, Gus (Claus Augustus), born in 1858, also had a partnership in many of the family holdings, including streetcar companies, the Commercial and Sugar Company of Hawaii, the brewery, and in six ranches and land companies, almost all of which today comprise Rio del Mar and Seascape, California. Emma, two years older than Rudolph, was estranged from her parents in 1896 after marrying Thomas Watson, a man twice her age, against their wishes. Angry at her father's reaction to her choice and having been raised a lady of gentility, she returned gifts he had made to her of the Punahou mansion in Hawaii and about $1.5 million in cash and moved permanently to England. Emma would not reestablish relations with her family until her husband died in 1904, after eight years of what seemed a happy marriage. Rudolph, being fourteen years younger than his next oldest brother, was yet to be born when the family business partnerships were created, which only added to his independence and determination. He would make it on his own.

The decade of the 1890s saw the Spreckels family riven by charges of embezzlement and a series of lawsuits that left Claus estranged from his two youngest sons. But when he suffered a stroke in 1903, like most men who experience a life-threatening illness, he must have taken stock of his life and his relationships with his children. Rudolph was by now prospering as a businessman and living the life of a respectable gentleman. He was well liked in San Francisco and said to be as charming as a sober prince. He had taken partnerships in banks and in insurance companies and built a tasteful mansion at 1000 Pacific Street as well as a country estate called Sobre Vista, in Sonoma County, just down the road from where Jack London lived and wrote. He had become everything that Claus approved of. So the vindictive father, when faced with his own mortality, decided to permit a reconciliation with his youngest son.

His parents' home suffered very little damage from the quake, and Rudolph now found them both shaken but well. His relationship with his father remained awkward, and after finding their staff in place and organized he continued down Van Ness. He was looking for his friends, any of them, for he was determined to assemble a vigilante committee as he expected widespread looting to occur in the chaos left by the earthquake. Like General Funston, he was primarily concerned with guarding the property of the victims and the helpless. That thought quickly abated because Rudolph saw all around him people in need of immediate help.

As the morning wore on he offered assistance in whatever way he could until at around eleven o'clock he returned his carriage to his carriage house, certain that he did not want it commandeered by the authorities. He anticipated needing transportation later in the day when he might have to transport his wife, family, and staff. He did not stay for lunch for he felt, as the day wore on and the smoke plumes grew larger and larger, an ever-greater urgency to get to his office and protect the important papers he had in his safe.

CHAPTER 35

Captain Arthur Welch had slept through the third alarm that Jack Murray and the men of Engine 1 had worked earlier, because Engine 7, at Seventeenth Street and Albion, was located far enough away from that fire that his company was not listed until the fourth alarm.

But just about dawn, he had been thrown out of his sleep by a great and sudden jerk, when it seemed that the whole of the firehouse shifted. The thin brass firehouse pole was still vibrating as the captain and the men of Engine 7 slid down to the apparatus floor. Their first thought was to calm the horses, but the firemen quickly noticed that the building seemed off kilter, and even the stall doors were jammed. They ran to the front of the firehouse and pulled with all their combined might to get the thick and heavy doors of the firehouse to slide open, but they were so resistant that it seemed as if they had been soldered to the floor. As the horses slammed themselves against their stalls and reared as much as they could in their confined spaces, the firemen went to retrieve tools from the basket beneath the hose wagon and managed to force open the side door of the firehouse, to an alleyway off Guerrero Street. Just then a citizen came running up the street warning of a fire nearby on Dolores.

Artie Welch had dedicated his entire life to the fire department, and all of his instincts told him that in this situation, in a major shake, every second was going to count if the company was to save any jeopardized lives. He immediately saw that it would take a prohibitive amount of time to ax through the apparatus doors to get the horses and the LaFrance steamer out to the street and instead immediately led his men in a run west, where they found civilians pointing to smoke rising several blocks up Dolores. They raced to that location, arriving just as Engine 19 rolled in, with Chief William Waters trailing behind. It was not a large fire, involving only a couple of rooms, and started no doubt like all the earthquake fires—with a fallen candle, a tipped benzene heater, a broken gas pipe, or a shifted flue. They all searched for water, but found the hydrants dry. As other fire companies arrived to help, the chief ordered Engine 7 back to their firehouse so they could extricate their apparatus and horses to complete their company.

It took them some time to chop through the doors of the firehouse, but as soon as they had freed the steamer a citizen came to report that the Valencia Hotel had collapsed, just one block in the opposite direction from Dolores. The men of Engine 7 mounted their steamer and hose wagon and drove directly there. Much of the building had sunk into the ground, and what was visible was badly crumbled. The top floor was sticking out into Valencia Street like a huge freight car thrown over the wreckage. A small group of men and women were working their way over the heap of wood, shingles, and glass trying to pull people from their entrapment. Captain Welch, a shovel in one hand and a pickax in the other, climbed to the top of the pile, where he was surprised to see former mayor James Phelan, shirtsleeves rolled up, balancing on a fallen beam, helping a man out of a hole. The men of Engine 7 followed, each carrying a heavy tool, and together they quickly cleared as much rubble as they could, until they found body after body, all annihilated by the crush of the heavy wood. They placed eight bodies in blanket-covered stretchers, and by the end of the day, forty were discovered dead in this building alone.

Earthquakes and the fires that inevitably followed in their wake had always been a topic in the drill periods for the San Francisco firemen, but not one of them had been on the job when the last big and damaging earthquake hit the city on October 21, 1868. That calamity in the Bay City killed five and brought down many buildings, but no story or drill period had prepared these men for what they were now enduring.

Captain Charles Cullen and the men of Engine 6 were facing a trying dilemma—a situation that was the reverse of Engine 7's, with its trapped LaFrance steamer. Their engine house at 6th and Folsom was still draped in purple bunting and the men were still reeling from the terrible choking deaths of their members Captain Dakin and T. J. Hennessey, some ten weeks before. The victims were from close-by companies Engine 4 and Engine 22, and so were well known to the men of Engine 6. It takes a long time for line-of-duty deaths to become accepted and then woven into the history of any fire department.

Engine 6's firehouse shook so much during the quake that the floor rose beneath their Clapp & Jones double-gated steam engine, lifting it from its wheels, and pinning the steamer by its own weight. At the same time, the doors of the firehouse flew open, as did the corral gate to the stalls at its rear, and the company's five horses ran wild eyed into the streets. The firemen, though, wasted no energy running after them. When the horses settled down from the shock of the quake, they would realize there was nowhere to go and, like mail pigeons, they would return. The firemen, however, had to get their steamer out of the firehouse, for even without horses, they could at least pull it to a nearby hydrant. With extraordinary effort they shoved and pulled the steamer over the broken stones of the flooring until they had brought it into the middle of the street, a little safer from possible falling rubble in whatever aftershocks might come. There they heard many cries of "Fire! Fire!" While most buildings they could see were still standing, the buildings to both the right and left of the firehouse were both collapsed and burning.

At that time, if a fireman lived no more than one block from the firehouse, he was occasionally allowed, with the permission of his captain, to spend a night sleeping at home with his family. All firemen had bells installed in their homes so they could hear the alarms as they came in, which would enable them to get back to the firehouse in time to catch the rig as it turned out for an alarm. The night before the quake, Hoseman John T. Titus had asked to sleep at home with his wife and children and was given permission by Captain Cullen. Titus lived right next door, so the request was easily granted.

The first thing the firemen saw as they rushed to the neighboring structures was the hand of a child sticking up through the rubble. They quickly dug her out, and discovered it was Titus's daughter. Working quickly they freed two other children and finally Titus and his wife. It was cause for some small jubilation in the disorder.

As some of the crowd helped them pull their steam engine to a hydrant, others appeared and reported the collapse of and fire at the Brunswick Hotel, just three blocks down at 6th at Howard. One hundred fifty people were trapped in that building, and the fire was keeping potential rescuers away. They needed the firemen, and they needed water. And just one block away, at 6th and Folsom, the Corona House Hotel had collapsed and was in flames. Forty people here lost their lives as well.

The firemen rushed to the sites of both calamities, but the water came only in a trickle, and most of the trapped were burned to death. It was a gruesome scene as the firemen were turned back, helpless without water or adequate manpower, as the victims screamed louder and the fires grew larger. Captain Cullen and his men did manage to pull two men from the top of the rubble of the Corona before the encroaching flames became too intense and then had to rush to save their steam engine. As Cullen recalled, "With wet sacks around our heads we assisted our engineer to uncouple his apparatus and pull the engine by hand along Folsom Street."

CHAPTER 36

South of Market proved to be the worst hit area, given that the ground there was flat and mostly alluvial. When the quake struck, many of the buildings shifted or simply sank, anywhere from inches to twenty feet. This sinking was caused by liquefaction, a process that dramatically affects alluvial soil in an earthquake, particularly the kind of sandy soil on which most buildings had been built South of Market. If you fill a coffee can with sand and then wet the sand just short of saturation you will see the process of liquefaction when you slam the can flat down against the tabletop several times. The sand will sink lower as it compacts, and the water will rise to the surface. The soil in San Francisco was solid enough to provide stable support for building two- and three-story homes, but it retained much moisture, which rose as the ground shook in the violence of the earthquake. In fact, less than a third of the buildings in this area of the Mission district were still vertical, for most had leaned over, twisted, or fallen to the ground. A Catholic priest, Father Ralph Hunt, recounted one particularly poignant scene he witnessed as he rushed to St. Vincent's School to save whatever he could. The fire had already come up this far from the Chinese laundry and was spreading quickly. The firemen were all about before him, laying hose and playing their meager streams against the fury of the building conflagration. At Mission and Third, a building had fallen, and he could see bodies caught in the rubble. Then he noted one man, his head free but the rest of his body trapped within the tons of debris.

"Don't leave me here to die like this," he pleaded again and again. Captain Thomas Duke of the San Francisco Police Department reported that he saw a large middle-aged man step forward into the wind of the smoke and heat. He spoke to the helpless victim for a minute or so, his hand held steady at the back of the man's head. The jumping flames of the fire were now just yards away. He rose, drew a pistol from his pocket, and took aim at the victim, killing him with a single shot.

The man then asked a police officer where he could find the mayor. He was taken then to the hall of justice and brought before Mayor Schmitz, who made time for him. The mayor listened to the man's account of the trapped and doomed victim and considered his disconsolate and distressed bearing. The mayor sent the man away and commented to others that he had "commended him for his humane act."

CHAPTER 37

The streets quickly filled with refugees. A businessman named G. A. Raymond was staying at the Palace Hotel. He had brought a large amount of cash, six hundred dollars, which he had placed beneath his pillow as he slept through the night. He jumped from his bed at the start of the shaking, stuffed the money into his pockets, and made for the street. "We must get to the ferry," he heard someone say, and he began to follow the crowds up Market Street. Hundreds shuffled along the street, carrying, pulling, and pushing their possessions. For the most part they were orderly and determined, but there was an occasional desperate cry from someone unable to cope with leaving the accumulations of a lifetime behind. At the waterfront, Raymond reported, he found just one boat, and there was pandemonium as people tried to board, elbowing and fighting their way forward. A crowd of more than ten thousand had by now gathered at the water's edge and was growing. People began fighting like wildcats to board the boat, and clothes were torn from the people in front. Some women fainted, and no water was available anywhere to revive them. Some men lost their reason altogether, and one banged his head against an iron pillar, crying out, "This city must be saved."

Raymond did not remember how he had made it onto the boat.

CHAPTER 38

Sometime after noon that first day, Ensign Wallace Bertholf reported to Lieutenant Freeman in full uniform and announced that he was coming off shore leave. It was a formality and, sensing the incongruity amid the heat and smoke all around them, Freeman smiled and passed the hose he was holding into Bertholf's hands. The lieutenant then moved down the line of seamen and firemen who had their streams aimed at several burning buildings on the far side of East Street (now known as the Embarcadero) near Howard. He was cheering his men forward and also keeping an eye out for any additional military men or police officers he might conscript for his efforts.

As he directed the operations, a soldier came up to him, a messenger from General Funston. Freeman was ordered to take the *Preble* down to Monterey, where he was to deliver a message to Colonel Marion Maus, the commander of the 22nd Infantry. (The soldiers of the 22nd had been divided between their bases in Monterey and Angel Island.) Funston wanted Maus to bring his unit back to San Francisco on the *Preble,* augmenting Funston's own forces by several hundred men.

Freeman then made another decision that could have ended his career. Uncomfortable with leaving his men without a leader at the fire line, he decided to send a warrant officer with the *Preble* to take the message to Maus while he, Bertholf, and Pond remained with the firefighting effort. To disobey an order from a superior officer, especially a general, is against all naval regulations, and such a violation would have been viewed in an especially critical light in Rear Admiral McCalla's eyes. But Freeman realized as he surveyed the wharf area and east to the railroad depots and yards that the fire kept spreading and the situation grew more desperate with every minute. He had so few men and believed that the firefighting had to continue without relief or interruption. Freeman knew that no other course could have been considered.

Though events would prove Freeman right, his decision would not be viewed by General Funston as exceedingly courageous. He had, after all, refused to execute an order. The fact that Freeman had never been singled out for praise by the army might also have been based on its natural reluctance to appear to have been dependent upon the assistance of another service. Funston's acknowledgments of the navy's contribution to saving the waterfront and parts of San Francisco's outer districts would be limited to a few sentences in his official report. In each instance he cites that aid "offered" came either from Admiral McCalla or Admiral Casper Goodrich. There is no doubt that the navy did a much more consequential job than the army in fighting and containing the fire. And though the army distinguished itself in so many ways during the recovery period, it can be said that its policy of spontaneous dynamiting of buildings caused a bad situation to become ruinous.

Of Freeman's heroic efforts, Funston would write and say nothing, though he was fully cognizant of the naval lieutenant's actions.

The *Preble* made the journey to Monterey in four hours, probably breaking all speed records up to then for a sea journey between the two cities. When the *Preble* returned with the soldiers, she was then sent from Fort Mason to Mare Island to pick up a contingent of marines under the command of Lieutenant Colonel Karmany. Lieutenant Freeman and the navy men were puzzled that Admiral McCalla had complied with the request for these marines to assist the army, for they were certainly much needed by the navy and could have been put to good work fighting the fires on the wharves. In the meantime, the Pacific Fleet, under the command of Admiral Goodrich on the USS *Chicago*, was making way for San Francisco from its position in San Diego, but would not arrive until the following day. It was a matter of pure luck that Admiral Goodrich had received an experimental wireless dispatch informing him of San Francisco's plight.

By late Wednesday afternoon only Freeman, Pond, Bertholf, and the sixty-six sailors of the *Perry,* aided by a few marines and firefighters, stood between the great wall of flame before them and the utter destruction of the piers, wharfs, and terminals.

CHAPTER 39

When word of the San Francisco earthquake reached New York late that Wednesday morning of April 18, William Randolph Hearst remarked that earthquakes happen all the time in San Francisco and that there was no need for him to leave his *New York American* to travel to the West Coast immediately. The railroad magnate E. H. Harriman, in contrast, made ready his private train to leave at dawn on Thursday and instructed his engineer to make a record trip across the country to the afflicted city. He had already established several notable travel records. Just the year before he had set the west-to-east record, with one of the country's most important personages on board—the daughter of the president. Alice Roosevelt had been on the Harriman steamer *Siberia* when Harriman set a record—crossing the Pacific from Tokyo to San Francisco. Thinking it then appropriate to match that achievement, this one overland from San Francisco to New York, he ordered his engineer to push the lever all the way forward. The press made much of this effort, and when luridly dramatic stories of a death-defying transcontinental race appeared in the newspapers, Harriman received a telegram from President Roosevelt regarding his daughter's safety. Harriman sent a telegram in return: "You run the country. I'll run the railroad."

Harriman, born in 1848, was the son of a failed entrepreneur and Episcopal priest and believed fervently that God gave man dominion over the earth and everything on it, a belief that motivated his passionate acquisition of railroads, steamship companies, real estate, and money. He had, at fourteen, the good fortune to head to Wall Street when the Civil War was beginning to prime the motors of the country's businesses; to get into the railroad industry in the 1880s, when there were more and more manufacturers clamoring to move more and more goods across the country; and to realize many opportunities in the country's "merger mania" in the last decade of the century. However, his accumulation of property and subsequent concentration of wealth led to his being known as an acquisitive octopus during the tenure of the nation's first trust buster, Theodore Roosevelt.

In the early 1890s, Harriman and Roosevelt had become good friends when they discovered a common interest in land preservation. The famous environmentalist of that time, John Muir, had asked Harriman's help in securing the land of the Yosemite Valley State Park for Yosemite National Park. This land had been given to the State of California by Abraham Lincoln in 1864, in an "inviolable public trust," a grant that may be the nation's first major effort at scenic preservation (and that perhaps makes Abraham Lincoln the true father of American environmentalism). In those years, though, there was great resistance in California in preserving the lands and the extraordinary giant sequoias of the Yosemite Valley. The federal government had created the Yosemite National Park in 1890, but the surrounding Yosemite Valley State Park remained under the control of the State of California, whose politicians wanted to reduce its boundaries to open the land for development. John Muir and 181 other like-minded individuals created the Sierra Club in 1892 to fight any attempt to reduce the size of the Yosemite Valley land trust, and their efforts were successful, giving the Sierra Club an instant reputation as a formidable political foe. But sometimes it would take more than nature-inspired advocacy to protect the environment, as Muir came to realize.

By the turn of the last century the Sierra Club had developed a campaign to return the state park lands to the federal government, believing the land would be better protected out of the control of the California legislature. Recession was favored by most citizens of the day, but the politicians balked, mostly because two powerful people—State Senator John Curtin, who owned a cattle ranch contiguous to the land in question, and William Randolph Hearst, for reasons of cronyism and the value the controversy would have in selling papers—were against the proposal.

In the face of Curtin and Hearst's opposition, John Muir turned to his new friend Harriman, whom he had met on a scientific expedition to Alaska just a few years earlier where a friendship in mutual admiration began. Harriman gave Muir access to the political lobby of the Southern Pacific Railroad in Sacramento, an organization of elbow-twisting professionals more powerful than any newspaper, and legislators were induced to speak out against the bill for their own political protection, but then to finally vote for it.

Harriman also contacted his old friend Teddy Roosevelt to secure his help, which Roosevelt eventually delivered. By 1904, however, Harriman had put together a large number of railroads in his Northern Securities Company, including the Northern Pacific, the Great Burlington, and the Great Northern, not to mention his acquisition in 1900 of the Southern Pacific. Roosevelt, convinced that prodigious concentration of capital decreased the possibility of healthy competition, brought an antitrust case against Harriman, forcing him to break up the Northern Securities Company.

The antitrust case had saddled Harriman with the reputation of being a cold-eyed, gilded-age despot—an ill-deserved definition for a man who had created the Harriman State Park in New York, the enduring financial firm of Brown Brothers Harriman, and the founding organization (the Tompkins Square Boys Club) for what is today the Boys and Girls Clubs of America. He was also a nurturing and attentive father whose son, W. Averell, went on to become one of the nation's most important ambassadors and foreign policy figures since Benjamin Franklin and whose daughter, Mary, convinced seventy-nine other debutantes in 1901 to forgo the hugely expensive dinner dance that is part of the coming-out ritual in America and to instead use the money to create the Junior League.

In 1905, Harriman and Muir were finally rewarded for their determination and persistence when the land of the Yosemite State Park was returned to the U.S. Congress for inclusion in the growing national treasure that was, and is, its park system.

Harriman left New York on April 19 and set a new east-to-west record racing to the tragedy, traveling across the country in just over three days and arriving in Oakland on Sunday. (On his return trip seventeen days later, however, he reached New York in seventy-one hours, averaging 50 m.p.h. and at times traveling as fast as 85 m.p.h.—a record that held for many years.) En route to San Francisco Harriman demanded updated information by telegraph at every water stop. He also telegraphed ahead to make certain that his train was loaded with provisions for the relief of the homeless, spending more than eighteen thousand dollars for foodstuffs alone that he gathered along the way.

His one great concern was that the city would be able to save the railroad sheds and the tracks, for they would be necessary to move sufficient people and materials for rebuilding. San Francisco, he realized, had to keep functioning without interruption if it was to survive this great catastrophe. Indeed, Lieutenant Freeman had come to much the same conclusion, and in the weeks ahead, the railroad sheds and depot would indeed become the most important center of activity in the reconstruction effort.

When he realized that hundreds of thousands of people were being forced to flee the city, Harriman immediately made his trains and shipping lines available free of charge for anyone who was in need of transportation, and more than 270,000 people availed themselves of that offer. On Thursday alone, 1,074 carloads of people left the city, and many thousands more escaped on Harriman's ships and ferries. On most of these trips the refugees were given free meals as well. Harriman made the Southern Pacific's hospital at 14th and Mission available at no cost for any medical attention that might be required. On the first day 250 victims were brought there. Thirteen died, but the others were then taken to San Mateo as the fire reached the hospital, and then on to the Southern Pacific hospital in Sacramento. Harriman's doctors and nurses worked for six days without a solid night's sleep, resting on stretchers and hard chairs.

Harriman was the first major personage to reach the city to offer assistance, and he was to stay the longest, not leaving until May 5. He worked tirelessly with the city's leadership, offering his advice, money, and relationships to relieve the suffering of San Franciscans, living all the while in the moving mansion that was his railroad car. Like most important doers in American history, Harriman had a firm belief in his own capabilities, and it was said that his self-confidence supported and inspired the work of those around him. He seemed to bring out the very best in people. Harriman's generosity set a precedent for a new American corporate responsibility in the face of catastrophe, a philanthropic rule of thumb that says: Those that can, should. One newspaper wrote, "The Harriman monopoly, if it be so called, has proven itself in the hour of need the strongest and most faithful friend that the city of San Francisco could have had."

CHAPTER 40

At the corner of 17th Street and Valencia Avenue, just a few blocks from where Rudolph Spreckels's brother John had his home, lived the former three-term (1896–1902) mayor of San Francisco, James Duval Phelan. He was sleeping in the same room in which his mother and father had died peaceably some years before, and when the violent shaking of the earthquake began he ran to the doorframe for safety just as the large chandelier broke from the ceiling, swinging wildly and dangling from a thin wire support. He hurried across the hall to see to the safety of his much beloved sister Mollie. The house was still shaking as they looked out the window and saw that the chimneys of their large Victorian had fallen from the roof into the garden. Phelan quickly dressed and went to the gas line connection at the street and turned it off, giving instructions to the servants that no cooking was to be attempted anywhere in the house. He knew immediately that fire would be a consequence of this severe a quake.

The street was pandemonium, with one dazed man covered by nothing but a red tablecloth. Looking north, Phelan could see the smoke that was already billowing from the many blazes that ran toward the Ferry Building, about a mile away. The sight reminded him it was only a few years before, in 1900 when he was mayor, he had insisted on an image of the phoenix be included in the design for the city flag that he proposed. San Francisco had survived six major conflagrations since 1850, most notably the great fire of 1852 (which almost leveled the city), and like the phoenix it had risen from the ashes each time. Now, there would have to be more rebuilding, Phelan must have thought, but San Franciscans were always ready to meet whatever was presented to them, to meet it with the fortitude and courage implied by the strength of iron as it was inscribed in Spanish on the flag, just below the phoenix: "Gold in peace. Iron in war."

Down Valencia Street, at 18th Street, he saw that the Valencia Hotel had been nearly swallowed up whole. Just two of its five stories remained above the ground, and these had pitched forward onto the sidewalk. A police officer who had been at the hotel at the time of the earthquake, Lieutenant Powell, would write, "Suddenly, Valencia Street not only began to dance . . . and roll in waves like a rough sea in a squall, but it sank in places and then vomited up its car tracks and the tunnels that carried the cables. These lifted themselves out of the pavement, and bent and snapped."

At the same time that many of the hotel's 120 guests were buried in their rooms beneath the ground, a water main had burst and filled whatever voids existed, drowning the hapless and trapped victims. Except for a few who were in their rooms on the top floor, not many survived the collapse.

A crowd of people had already gathered there, responding to the screams of those trapped within. Phelan ran to his stable to gather axes and saws and enlist his servants, and, piling them all into his carriage, made his way back to the Valencia. He was two days away from his forty-fifth birthday, and though not a tall man, he was an imposing man in superior physical condition. He quickly climbed to the top of the building and began to dig through the rubble of the fallen structure. Everyone there that morning, he later reported, worked "with vigor and intelligence" in getting those few survivors loose from their entrapment. Phelan then gave over his carriage so that the injured could be taken to the Southern Pacific Railroad's hospital at 14th Street. His own death would be incorrectly reported in the San Jose newspapers the following morning, because an eyewitness had seen "an apparently dead person" in the back of his carriage.

When his carriage returned, Phelan proceeded immediately to his office at the Phelan Building on Market Street, seeing the ruins of city hall as he passed north. Ever the classicist, he described the columns that withstood the collapse of the huge and famous building as reminding him of the Temple of Dioscuri in the Forum in Rome. He also passed a group of Royal Arch Masons all wearing their red fezes, heading for their temple at Montgomery and Post, determined to attend the second-day ceremonies of their annual convention. These men, whose calendar begins with the building in 530 B.C. of the second Temple of Jerusalem, recorded the destruction by fire of that structure several hours later as of the date April 18, anno 2436.

As he arrived at his own building just across Market Street from the eighteen-story Spreckels Building and a block down from the Palace Hotel, Phelan instructed his carriage driver to search for Jimmy Mountford, his chauffeur. Before entering Phelan met General Funston commanding a detachment of troops, their bayonets fixed, lined up where Market and Grant streets cross. (Curiously, Phelan did not mention this meeting in an account of the fire he wrote ten years later.) Together the two men, finding the elevators inoperable, walked up the stairs. Funston's offices were on the fourth floor, where he shared headquarters with the commander of the Pacific Division, Major General Adolphus Greely. Funston proceeded another flight up to Phelan's office, and as they surveyed the mess of books and bric-a-brac covered with plaster and scattered across the floor there was suddenly another tremble of the earth, shaking the entire building. "This is no place for me," Funston exclaimed and quickly retreated to the safety of the open air.

Phelan hastily searched through his desk until he found his most important papers: the articles of incorporation for the new company he had created with Rudolph Spreckels and that had just been approved the day before, to create a new railway system in San Francisco—a railway important to Phelan's City Beautiful plan.

Then with the help of janitors, he carried some artwork—pictures by Benjamin Constant and E. L. Weeks and a sculpture by Rodin that he had purchased directly from the artist—other important documents, and the family scrapbooks, and instructed his driver to keep these valuable possessions in the middle of Union Square. But because of the ensuing events of the next few days, he was to forget completely about them, his carriage driver, and his order to preserve them at all costs. Fortunately, his driver stuck by them and ensured their safety.

After leaving his office Phelan walked a few blocks to the First National Bank (formally the First National Gold Bank) of San Francisco where he was a director, and Rudolph Spreckels was president. There he was met by socialite and businessman, J. Downey Harvey, who had a message for him from the mayor, requesting him to come to an emergency meeting of civic leaders to create a relief committee—a group of men that would be known in those first few days as the Citizens' Committee of Fifty, and later as the Committee of Forty. It would fall to them to manage the affairs of a catastrophe.

CHAPTER 41

"Police work," recently retired inspector of the San Francisco Police Department Kevin Mullin wrote, "is not all heroics. Police officers help people with all the mundane little problems encountered in daily life." Because this was as true in 1906 as it is today, very little has been written about the work of the Bay City's finest during the earthquake and fire. Not surprising, the police force worked long and hard hours, assisting, carrying, directing, transporting—everyday duties as important in a crisis as during the daily hustle and bustle of city life. The story of Max Fenner, however, stands out. He was the only police officer killed during the emergency, and his death was reported as heroic.

Working the midnight tour, Fenner had an easy night, though one of the two prostitutes he had arrested just after midnight gave him a somewhat rough time, and he remembered lifting her completely off her feet by her elbows to get her to move. Often called the Hercules of the police department, Fenner was tall and very mighty, but at 5:12 A.M. his great strength offered no defense against the falling building that crushed him, just in front of the Oriental Bar and Café at 138 Mason Street, where he had been talking with the manager. The early reports of that day all described a police officer being crushed by a collapsing building in the course of duty. No mention was made of anyone else's being involved, except for the manager, who had pressed himself against the wall of 138 Mason and who had only been hit on the shoulder by falling bricks. It was widely reported the following day in newspapers throughout the country, however, that Fenner had seen a woman coming out of a building across the street. He called out to her and ran frantically to save her from the falling masonry. She heard his warnings and drew back into the building, but Fenner could not outrun the collapse of the wall. Another woman, "a very lovely girl," an unnamed policeman said, had plummeted from the fifth floor and lay cold on the rocks above the police officer as passersby attempted to dig him out. Heroes can be made by dramatic action as much as by commitment to duty, and in the case of Max Fenner there is convincing evidence to say that he was heroic on both counts.

Sergeant Jesse B. Cook, who would go on to become the city's chief of police, was also working the midnight shift. He loved his job and unlike some other police officers he did not nap in an out-of-sight resting place during the graveyard tour. On April 18, in the early-morning hours before the earthquake, he was doing a patrolman's work, going from store to store, factory to factory, checking the security and community peace.

At 5:12 A.M., Sergeant Cook was standing in front of the Levy Produce Company in the wholesale produce district, just north of Market Street and two blocks from the waterfront at Washington and Davis. He was speaking with Sidney Levy, the proprietor's son. Because of the early hour few people were on the streets, but those who ran the city's wholesale markets—the butchers, bakers, fishmongers, and fruit and vegetable dealers—were already well into their working day. Amid the market bustle, Cook's attention turned to a horse that had been rigged to a wagon in front of Levy's. The animal was shaking his head, neighing, and stamping its forefeet over and over. "Something is making that horse nervous," Cook said, and just then he heard a distant rumble, "deep and terrible," rushing toward them. As it came closer Cook realized it was an earthquake, and being familiar with earthquakes from his ten years of walking the beat, he said, "It's a dandy."

Looking up toward Russian Hill he saw the whole street before him "undulating . . . as if the waves of the ocean were coming towards me. . . . The houses were nodding and bowing to each other across the thoroughfare." A rift opened in the middle of the street, several feet wide, and with great effort Cook jumped over it, running to a one-story building for some protection, away from the three- and four-story buildings around him. He noticed that the rift had begun to fill with water. He stood in the archway for a couple of seconds, and then a second shock came that was much "fiercer and sharper." The ground suddenly began moving and twisting, first one way then another.

Across the street three men had begun to run out of Bodwell Brothers Commission Merchants, and Cook yelled to them to remain where they were, but only a porter stood his ground. The other two, Mr. Bodwell

Lieutenant Commander Frederick Freeman (*left*), the forgotten hero who saved the wharves and railroad sheds of San Francisco. Freeman's loyal assistant, Ensign John Pond (*above*), took one of the early photographs of the fire.
(Courtesy of the Pond Family Collection)

The U.S. torpedo boat destroyer *Preble,* which carried Freeman, his sailors, and a cadre of doctors from the Mare Island Naval Base to the burning city.
(Courtesy of the Pond Family Collection)

Eugene Schmitz, an orchestra leader who was elected mayor of San Francisco on the first successful labor union ticket in America. *(Courtesy of the San Francisco History Center, San Francisco Public Library)*

Abraham Ruef, the cultured reformer who went on to form one of the most corrupt mayoralties in American history. *(Courtesy of the San Francisco History Center, San Francisco Public Library)*

Brigadier General Frederick Funston, who, in taking full charge of the San Francisco conflagration, made a number of bad decisions that led to the ruination of much of the city and left two hundred thousand of its citizens homeless. *(Courtesy of the Kansas State Historical Society)*

Engine Co. 1, San Francisco Fire Department. Jack Murray (*fourth from left*), the patriarch of a great firefighting family, spent days battling the blazes.
(Courtesy of the Murray Family Collection)

One of a long line of troops marching into San Francisco, their bayonets fixed.
(Source unknown)

A row of wooden houses destabilized by the 8.0 earthquake. Of the twenty-eight thousand buildings ultimately destroyed in the catastrophe, only 2 percent were attributed to the quake itself, while fire took the rest.
(Courtesy of the Bancroft Library, University of California, Berkeley)

A crowd of men watch as rescue attempts are made in the rubble of a collapsed structure. Firemen often enlisted the help of civilians before the army mandated evacuation.
(Courtesy of the Bancroft Library, University of California, Berkeley)

A view of the fire captured early on the day of the quake in front of the Italianate-domed Call Building, owned by the Spreckels family. Ultimately, every building in this photograph was destroyed.
(Courtesy of the Bancroft Library, University of California, Berkeley)

Evidence of a slow-burning fire on the nearly windless first day of the fire. Onlookers in all of the extant photographs show no signs of sensing immediate danger.
(Courtesy of the Bancroft Library, University of California, Berkeley)

A calm bay fronts the threatened Ferry Building at the top of Market Street, in a photo taken late on the first day as the fire burned through the business district.
(Courtesy of the Bancroft Library, University of California, Berkeley)

A photograph taken by Ensign John Pond as he steered the tugboat *Active* into San Francisco Harbor. *(Courtesy of the Pond Family Collection)*

An engine company furiously fights a fire downtown with water tapped from a cistern. *(Courtesy of the Virtual Museum of the City of San Francisco)*

A view from beneath an immense smoke cloud, as seen from across the bay.
(Courtesy of the Bancroft Library, University of California, Berkeley)

Arnold Genthe took the most well-known and best of the thousands of photographs of the fire. Here, he captured two unlikely smiles in the face of encroaching disaster.
(Courtesy of the Virtual Museum of the City of San Francisco)

Residents skirt the fire in a dangerous last attempt to save their property.
(Courtesy of the Virtual Museum of the City of San Francisco)

Many evacuees—the famous tenor Enrico Caruso among them—believed they could wait out the fire and gathered in Union Square.
(Courtesy of the Virtual Museum of the City of San Francisco)

Others sought any open space in the city to save themselves and their meager belongings.
(Courtesy of the Virtual Museum of the City of San Francisco)

Citizens begin to assemble and watch idly as the flames creep toward the famous Call Building, owned by the Spreckels family. *(Source unknown)*

The crowd of onlookers has grown larger as the Call Building is engulfed in fire. *(Courtesy of the Virtual Museum of the City of San Francisco)*

The burned-out shells of the business district.
(Source unknown)

A view down Market Street from the top of the Ferry Building.
(Courtesy of the Bancroft Library, University of California, Berkeley)

The collapse of some buildings, such as the toppling of the dome of the hall of justice, was mistakenly attributed to the earthquake rather than to the fire.
(Courtesy of the Bancroft Library, University of California, Berkeley)

The interior of the Hall of Records. Most official records in the city were saved, though newspapers reported them lost.
(Courtesy of the Bancroft Library, University of California, Berkeley)

The wedding cake Call Building was ultimately rebuilt and continues in use today.
(Courtesy of the Bancroft Library, University of California, Berkeley)

Abe Ruef's domed office building was also rebuilt, and when Ruef was finally released from prison after his corruption conviction, he opened an office there.
(Courtesy of the Bancroft Library, University of California, Berkeley)

James D. Phelan, (*left*), former mayor and future U.S. senator, led the relief efforts, and with his friend and business partner Rudolph Spreckels, (*right*), dedicated himself to ending the city's notorious corruption. Spreckels and his wife were ostracized because of his determined reform efforts. *(Left, courtesy of the Bancroft Library, University of California, Berkeley)*

The team that finally defeated corruption in San Francisco. Left to right in this famous photograph are prosecutor Francis J. Heney; President Roosevelt's favorite Secret Service man, William Burns; Fremont Older, the editor of the *San Francisco Evening Bulletin;* and the man who contributed two hundred thousand dollars to launch the investigations, Rudolph Spreckels. *(Courtesy of the Bancroft Library, University of California, Berkeley)*

and his clerk, dashed forward just as the building fell, burying them under a ton of bricks. It was too late to save them, thought Cook, for they could never have survived that collapse.

All around, he saw buildings swaying, cornices and chimneys tumbling to the ground. Cook knew that his duty was to return to his police station four blocks away, and as he started in that direction he noticed that the rift in the street had vanished, and he walked over what was now simply a crack in the ground. It seemed stable enough, but as soon as a laden cart drove over it, the street collapsed completely.

When he reached the corner of Clay and Davis he saw what was to be the first recorded fire that morning—the Irvine–Stanton wholesale grocery warehouse, where smoke was pouring out of the windows. Not knowing, of course, of the fires that had already started in many other parts of the city, he decided to run to the fire station of Engine 12, at Commercial and Drumm. On his way there, the Weiland Brothers Building fell just as he was passing, the cascading bricks injuring his shoulder, but he kept on until he met the captain of Engine 12, Jerry Sullivan.

The interior of Engine 12's firehouse had partially collapsed, but the firefighters had fortunately pulled their horses and engine to the street with the first shaking. Standing idly in the street, they noticed a peculiar absence of alarms, not realizing that several rows of battery acid bottles in the department's dispatching office a mile away—the batteries that powered the telegraph system that ran through the city's firehouses—had been thrown to floor, and 556 of the total of 600 were smashed into pieces. Not one official alarm was ever sounded for the largest fire ever to be fought in a major city, and more general alarm fires arose in these immediate minutes after the earthquake than San Francisco's firefighters would normally fight in several years. The department was now without any dispatching capabilities whatsoever. So, with a critically injured Dennis Sullivan and a ruined dispatching center, the department's brains and voice were no longer available to formulate the strategy so vitally needed in the beginning minutes and consequent hours of the catastrophe.

Informed by Sergeant Cook of the Davis Street fire, the fire captain and men from Engine 12 sped to the scene, only to find as they attempted to set up a hose stream that the fire hydrants were completely out of commission. The quake had shifted the earth and ruptured the water mains in several places. One of these mains was almost totally destroyed because it followed the fault line itself for more than six miles, traveling from a reservoir twenty miles away. Other vital pipelines crisscrossed over the fault, through marshy and landfill property in many areas, which subjected them to stress and breakage, rendering the pipes useless. The fire, meanwhile, churned forward like a locomotive that would inevitably require more energy to stop with every second of power it built. The red of the flames was reflecting from the cobblestones of the street, and Captain Sullivan pulled back, cursing his bad luck beneath his breath as only a proven fireman can curse. There is no greater frustration for a firefighter than being forced to let a fire burn, unless it is to see a victim perish before being given the opportunity to be saved.

Just down the street was the big warehouse of the Armour Packing Company. It had gone afire, and as the firefighters turned their attention there to at least do a search operation, the front wall fell. Captain Sullivan shouted a warning, and the firefighters scrambled out of harm's way. But one, George Wells of Truck 1, was knocked over by a mass of falling bricks. As he screamed in agony, his fellow firefighters dug him out and carted him quickly away for medical care. He was eventually brought to the naval hospital on Goat Island, where a crushed leg was amputated. Though many injuries were sustained by firefighters during the great conflagration, Fireman Wells's leg was only one of two maladies that were officially recorded.

In the meantime, Captain George Brown and the men of Engine 2 responded to a significant fire in a drugstore on O'Farrell and Taylor, one that would surely grow to take the large apartment house above it. The firemen found the nearby hydrants had no water at all, and Captain Brown, seeing a construction site just next door, had his men gather bucketsful of sand from there, assisted by a number of civilians. Surely and steadily they extinguished the blaze by heaving the sand at the flames, illustrating the fact that with enough manpower and imagination, any cooling agent (the sand interfered with the oxygenation needed to sustain the heat level) could be used in a firefight.

Engine 2 was then told of the collapsed Geary Lodging House, to which they hastened and were "confronted by a most heartrending sight." The building was a giant pile of rubble, and many people were pinned and trapped beneath the debris. Again, there was no water, and the firemen immediately climbed to the top of the pile and worked valiantly pulling, lifting, and throwing the planks, roofing, and flooring that imprisoned victims. A crowd assembled to cheer the firefighters on, but the men could take no comfort in the applause because they understood they were in the midst of a tragedy. The rubble was burning, and there was no time for thinking about anything but extinguishing the fire before it reached the many who were buried deep before them.

As Captain Brown carried a victim down what remained of a stairway, the wood gave away and he and the victim fell through the splinters. In the second recorded injury of the day, Captain Brown dislocated the bones of his right foot, which was also sprained and dangerously swollen. The pain had to have been great, but the captain drew himself up and proceeded, limping, down the pile with the victim still in his arms. Brown then called out to the crowd, asking them to join in the rescue work, and both the men and women began to clear away the debris. He dispatched another group of civilians to look in the neighborhood for dirt or sand, and before long dirt appeared in canvas bags and on top of squares of linoleum. In this way they reached the seat of the fire and were able to extinguish it, little by little, smothering and beating out the flames.

Seeing that the emergency was under control, Captain Brown brought his men out, and with a bleeding, broken foot, he led them to the next fire they saw as they looked up Geary Street toward Third and Mission. There again, they searched vainly for water, hydrant after hydrant, finding each one dry. They would have to enlist civilians, fire by fire, as they had in the Geary Street collapse, if they were ever going to stop these blazes from becoming a conflagration.

CHAPTER 42

At 5:12 A.M. on Wednesday, as the ground shook, Jack Murray, dogged and beaten from the third-alarm fire just a few hours before, thought of his family in their home on the top of Telegraph Hill. It was a natural reaction, and no doubt all the city's firemen and policemen had a similar concern. But they were on duty, and their job had suddenly been transformed from a daily routine to a colossal confrontation.

In just a few minutes, Jack was in the bucket seat of the steamer, cracking his whip above the heads of his team. The men of Engine 1 were following on a hose wagon. A citizen had run to the firehouse sounding the alarm for a fire just south of Market at Steuart Street, two streets from the Ferry Building. As the horses galloped Jack saw a man waving to him, calling for help at a building collapse at the top of Geary Avenue. But the Steuart fire was his priority, though he did not realize how many fires were already beginning to swell at that very moment.

As he neared the burning building, Jack's arms pulsed with strain as he pulled the horses back. Before him was a three-story frame structure at 14 Steuart that housed Smith's Cash Store. Next door was the Patterson Express Company, just one block from the Southern Pacific Ferry to Oakland, Alameda, and Berkeley.

Jack brought the team as close as he could, which was next to a hydrant just off the corner of Steuart. But the alarm had been so delayed that the flames were already blowing out of the windows on the second floor, and extending into Brown's Store at number 40. The neighing and whinnying horses pranced in place. It was too hot and dangerous to hook up to that hydrant. There was another toward East Street (now Embarcadero), but Jack wisely decided to take the team to the water's edge, near the ferries, and draft directly from the bay. The steamer was capable of pumping salt water, though all the manuals advised against putting salt in the contraption. Still, the bay had an endless supply of water, and all they would need was a supply of freshwater to create the steam itself, which they could obtain if necessary from the ships tied to the docks. The other men of Engine 1 stretched the two-and-three-quarter-inch hose from the back of the hose wagon, connected it to the steamer, and pulled it in fifty-foot lengths the two blocks to the fire.

Just nine blocks from Jack Murray, the firemen of Engine 4 and Truck 1 were fighting the fire in the Chinese laundry just across from the firehouse at 676 Howard, breaking down doors, searching for occupants. Their captain, Robert Woods, was trying to confront the death of James O'Neil, thinking what meager words he might say to Mary O'Neil at the funeral. There were small children in the O'Neil house, and Mary would have to learn to live on her dead husband's pension of $135 per quarter, less than half of the $1,200 annual pay of a truckman.

Suddenly, though, he had a more immediate concern. "No water!" was being yelled through the street. "No water!"

Captain Woods sent his men to a cistern he knew was located at Folsom and 2nd, which could hold twenty thousand gallons. But the cistern had not been well maintained by the Department of Public Works, and the firemen found a tank that was little more than half filled. The fire in the laundry had by then spread to the north exposure, and the firefighters saw that it would burn building to building, north or south, whichever way the wind took it, wherever wood was exposed to its radiant heat. Even after Engine 10 and Engine 35 arrived there were no more than eighteen firemen at the scene, too small a number to begin to tear down the buildings in the fire's pathway in an attempt to stop it. At the least, with more men they would have been able to throw all the furnishings into the middle of the street and reduce the volume of fire, which would in turn reduce the amount and the temperature of radiant heat, enabling them to get closer to the fire itself.

Meanwhile, after setting the steamer at the water's edge, Jack Murray gave its operation over to the stoker and ran to join the men of Engine 1. The water stream that was being pumped from the bay proved to be adequate—maybe three hundred gallons a minute—but the fire before him had now engulfed three buildings. Captain Murphy kept pushing his men, but they realized as they tried to cool down one side of the fire that the other side was extending farther down the building line on Steuart. The street was filled with heavy black smoke that was banking down to the gutters, leaving the firemen choking and gasping. Murray took his "elephant's ears" out of one pocket and a small vial of mineral oil out of another. The "ears" consisted of two small and attached sponges for the mouth and nose, which Jack soaked with the mineral oil and then tied around his head, hoping that he would be able to keep from coughing. He then began pulling the heavy hose, now filled with water and weighing about ninety pounds per fifty-foot length. He could feel the black grime of the smoke building on his face, but the mineral oil on the sponges allowed him to breathe more easily as it filtered out some of the heavier particulates of smoke.

The firemen all knew they would have to work harder, cooling as much as they could from building to building, and at times running with the hose line, which was like running with the weight of a piano. But pulling their jackets back to get a draft of clean air that might be trapped near their shoulders, they kept on. Some took their shirts off, wet them in whatever groundwater they found, and wrapped the shirts around their faces and heads.

Toward the middle of the morning, Patrick Shaughnessy, the assistant chief engineer of the fire department, arrived in his wagon. Although it was difficult for him to direct his carriage horse through the smoke and rush of the street throngs, he dismissed it as just another frustration of a day filled with obstructions and interruptions. He became further agitated as he watched the firemen move the hose up and down the line of buildings, cooling the heat, killing the flames. He wanted them to finish their work with this fire because he needed them elsewhere, and quickly. The chief began yelling commands and words of inspiration and support. The sleepless firemen had now been working for five or more hours, and they began to feel the pain of the constant movement. If only they could establish a powerful water stream, set it down in one place, and sit still for a little while.

But it was not to be, for Chief Shaughnessy needed them two blocks south, at Beale and Market, just across from the original Spreckels Grocery Store on Pine, and his need was desperate. Finally, the fire began to darken down, the red of the flames no longer leaping up and out. Fortunately there was no wind. This fire could be called under control, but it was far from being completely extinguished.

Suddenly a group of sailors and marines appeared, pulling a hose from a tugboat sitting at a dock next to the large ferry piers. It was Freeman and his crew, and a saving grace, the chief must have thought. He would make good use of these stalwart military men.

Tradition and pride would in ordinary circumstances dictate that one fire company would never give up its position to another if it could help it. Chief Shaughnessy, however, had his own priorities and ordered the men of Engine 1 to leave this fire and rush to the Beale Street fire. There was no communication system to order a multiple-alarm fire, and even if there was, the thirty-eight engine and ten truck companies now scattered throughout the city were already attending to one of the more than sixty fires burning at that moment, and were probably fighting many other fires that had gone unreported. "In truth," Shaughnessy wrote, "there was no need of bells," for a raging fire could be seen from every firehouse door.

After helping to disconnect the hose lines Jack Murray drove the steamer to a fire hydrant in front of the Holbrook Building at the south corner of Beale and Market. The McNutt-Kahn Stationers Building at 306 Market was completely involved in fire, but Jack and the team braved the radiant heat, bending low and near to the ground as they connected to the hydrant. He turned the hydrant valve and a gush of water fed through the hose, a strong, powerful stream. The fire was already into the O'Brien Building next door and halfway down to Fremont, and pounding in Jack's mind as he ran with a hose to the front of the stationery store was the desperate need for additional manpower. Chief Shaughnessy attested that these valiant firefighters of Engine 1 alone did the work of a third-alarm assignment of fifteen companies. He also wrote, "Nothing less than a dozen powerful streams would have stopped this fire."

Chief Shaughnessy watched the vibrations of the steam engine. It was such a powerful machine—a far cry from the first hand-pumped engine that the SFFD bought in 1849, called the Van Buren because it had previously watered the lawn of the former president's country estate. A Clapp & Jones double-valve model, the steam engine could pump five hundred gallons per minute, if only the water would hold up. At least here they had the water.

The department's eight chemical companies, which normally extinguished 60 percent of San Francisco's fires, would prove to be of no use on this day. The chemical wagons were typically very effective in responding quickly to a fire and extinguishing it in its early stages. Just two firemen manned each wagon, and it was easy work to stretch the small one-inch hose. The unit carried two eighty-gallon tanks of water mixed with bicarbonate of soda. When pressure was needed the firefighters would pull a lever and dump in a load of sulfuric acid, which acted as a catalyst, creating a gas pressure of about 300 psi at the top of the tank, which pushed the water out of the bottom and finally through the quarter-inch tip of the nozzle. Used carefully, a single tank could last for five minutes and knock down an entire roomful of fire. But this morning, with so many fires quickly extending, the chemical wagons were of little use.

Seeing that the men of Engine 1 were fully engaged, the chief wished them luck and rode off to find Engine 4 and to inspect the fire that had been reported to him. At this point, so early in the day, Shaughnessy believed that the fires could be stopped. He had seen water flowing through the hoses at Steuart and at Beale, which gave him a sense of optimism. But his view was soon changed when he learned that the firemen of Engine 4 had not been able to find water to fight the Chinese laundry fire, and it had gone well out of control. Joined by Engine 10, they had found a second cistern at Folsom and 1st, but it, too, held less than twenty thousand gallons and ran dry in a matter of minutes. Engine 35 found a hundred-thousand-gallon cistern at 1st and Harrison, a block farther south, but the fire kept claiming one building after another, pushing toward them. Finally, in desperation, Engine 35 went to salt water, at Third and Mission Creek, about a dozen blocks from where Jack Murray had been drafting from the bay. It was here where they tried to hold a line to protect the sheds of the Southern Pacific Railroad until later in the day when they would be pulled away to assist in the business district fire.

As Chief Shaughnessy watched in utter frustration as the flames moved from building to building out from the Chinese laundry, an officer came to tell him of the particulars of Fireman O'Neil's death. The news shook him greatly because he knew O'Neil and O'Neil's family well.

In the midst of so many setbacks was an especially cruel irony: Just across the street, not far from the quarters of Engine 4, was the warehouse of the California Wine Association, and stored inside that building were two and a half million gallons of red wine.

By the turn of the twentieth century, the State of California was fast becoming an internationally important winemaking center. Several of the labels from Napa and Sonoma counties had won prestigious medals in European wine competitions, and the industry was growing rapidly. Since wine spoiled in the heat, the City of San Francisco had great appeal for winemakers because of its consistently cool climate. (All San Franciscans are aware of the sudden temperature changes from one side of the bay to the other—as much as ten or twelve degrees on a good day—caused by the configuration of the coastline, which funnels the cold ocean air right across the center of the city.) A group of these winemakers—

including Lachman & Jacobi, Kohler and Frohling, and B. Arnold & Co.—had formed the California Wine Association, which maintained a large warehouse of wine vats on Townsend Street. In fact, from that location, stretching in a northward arc through the produce district, the financial district, and North Beach—the very path of the evolving fire—the various companies maintained a series of storage facilities, and it is estimated that on that day in 1906 more than forty-five million gallons of wine were being held in these warehouse vats, barrels, and puncheons. Furthermore, canvas hoses and pumps were available in every location, used to pump wine from one end of a warehouse to the other. Local saloons also were holding more than five million gallons in bottles, jugs, and kegs. In a city whose saloons had been ordered closed, there was undoubtedly enough wine to extinguish the early fires.

CHAPTER 43

As the day wore on, the San Francisco Fire Department was challenged as never before. Fires were raging in the western addition, the vast area west of Van Ness, but because water was available there, for the most part the firemen got the upper hand. But this was not the case at Hays and Laguna, Buchanan and Hayes, Golden Gate and Buchanan, Fulton and Octavia, or at dozens of downtown fires, in the financial district and in North Beach, just up from where Jack Murray and the men of Engine 1 were watching the surge of their hose stream diminish to a trickle, and calling on unimagined energy to pull the now limply charged hose back from the heat of the flames. The water had given out; the ruptures in the piping system had gained the upper hand. The firemen were losing water, and they were losing the fight at Beale Street.

Chief Shaughnessy returned to the downtown fires and was met by Chief Dougherty. With Dennis Sullivan they had been known as the department's divine trinity, but now a leaf was gone from that shamrock, and the two chiefs huddled to talk out what Sullivan would have done had he been with them.

All fire chiefs know that in big fires they are inevitably confronted by chaos, and their responsibility is to try to slowly organize the chaos, little by little, until it is brought into the realm of what can be controlled. Generally, the chaos lies in an area that can be contained at its perimeters, but in San Francisco on this day there were hundreds of perimeters to the fires, hundreds of exposures to additional fuel in the form of wooden buildings. Added to this were the loss of the dispatching system, which resulted in a splintered and uncoordinated firefighting effort, and the lack of a coherent strategy.

The chiefs agreed that Sullivan had often discussed the Baltimore conflagration and declared, if ever there was a great fire in San Francisco, that it would be better to take a stand at a single position, rather than attempt to fight each fire as it arose. This tactic raised two obvious questions, though: When and where to take the stand? Many fire companies throughout San Francisco might even now be making headway or preparing to take a stand, and to pull them away from any given fire could be counterproductive, or even dangerous. Still, from one side of Market Street to the other was a 120-foot breadth, and Van Ness provided a wide span as well. These two streets not only provided natural firebreaks, but together they also neatly divided the city into quarters. Flying cinders the size of golf balls filled the sky as the fires grew, and any slight gust of wind sent them toward new sources of fuel. In this way, the fire had now jumped across the wide expanse of Market, and a concerted stand here might well have stopped the spread of the flames from entering the financial district.

Jack Murray and the men of Engine 1 were now joined on the north side of Market by the men of Engines 2, 28, and 31 who had been fighting a fire at Sansome Street. Half the firemen were going from block to block, searching for water, as others dropped a hard suction hose of a steamer down into a sewer hole and tried in vain to suck up the raw sewage and whatever water might come up with it.

Chief Shaughnessy, seeing the futility of the efforts all about him, ordered the firefighters to retreat west, toward North Beach. By now Shaughnessy was realizing that there did not seem any way to control the fire. Was dynamite the answer? Every experienced fire chief in America knew the positive consequence that could be attained from dynamiting buildings when the explosions were carefully planned within an overall tactic, and when the right type and amount of dynamite was used. Shaughnessy had begun to consider this approach, not yet aware that General Funston had already had dynamite delivered to the hall of justice and that Chief Dougherty had been present when it had been ordered.

CHAPTER 44

Many of the citizens of San Francisco determined to do whatever they could to assist those who needed help. If they were not refugees, they took to the streets to volunteer wherever it seemed they were needed. Among them was General Funston's wife, Eda, who walked the three dozen blocks from her home at Washington and Jones to the Letterman Hospital in the Presidio. She knew that it had been newly renovated and equipped, and would serve the medical needs of the crisis well.

Eda Funston cared for the helpless as they came in, preparing places for them and assigning doctors and nurses. After being on her feet for twelve hours, she went rummaging through the basements and storerooms of the Presidio buildings until she came across a great surprise: a long-forgotten supply of two thousand blankets, the most fundamental thing needed to comfort the refugees pouring in by the hundreds into the Presidio.

Many of the soldiers also had great hearts and worked themselves weary to help the people of San Francisco. Three weeks after the fire, one of them, Charles Rhoda of Company H at the Presidio, wrote to his sister:

> We were sent down town in the morning and went to work carrying out the dead and caring for the wounded, and now sister it was a terrible site. And [those] who could not be gotten out had to be left or burnt up within the buildings, which of course numbered into the hundreds. And to hear the moans of those poor people who were wounded and burned

was pitiful, and it is a wonder to me how the boys and Red Cross people stood it. . . .

I myself worked 48 hours without food or sleep. The first day 4 of us boys handled 460 people who had to be carried on litters. Could do nothing for themselves, and you carry them away anyhow. . . . I am glad that none of my people lived here or have to be among the suffering. . . . Thousands of people live in tents and have nothing [—] only what they wear, no money, no nothing only what Uncle Sam gives them. . . . What is going to become of them is what I want to know?

CHAPTER 45

At about 11:00 A.M. on Wednesday, Rudolph Spreckels was making his way laboriously to his office at 421 Market Street. Few people in the midst of the chaos paid him any attention, despite his height, imposing posture, and determined pace. Few stepped out of his way. Like his business partner and friend James Phelan, he wanted to save whatever papers he could before they were claimed by the fire. Spreckels at this time had many business interests and owned, independently or in partnership, sugar refineries, banks, land companies, and a gas and electric company. The stock certificates of these companies were important, certainly, as were the family papers, such as his mother's will and his own. Most critical were the certificates of incorporation of the Municipal Street Railways of San Francisco, which were filed just the morning before.

Rudolph had never seen the streets so crowded, and he knew the situation would become even more difficult, for he could plainly see the fires just three blocks down Market on Steuart. People will have to get by those fires to reach the waterfront if they are to find boats, he thought. The fires were behind him as well, flaming past city hall and across to Mission.

Finally reaching his office, he sent the staff home to look after their families. Searching through the papers neatly filed there he took the small amount of cash he found and shoved the documents he sought one after another into a briefcase. He did not realize how prescient this action was, because three quarters of all the safes in the city would fail to protect their contents, as the heat that surrounded them would grow so great that any paper within would wilt, curl, and disintegrate. He looked for the railways certificates, and he was relieved upon finding them. This company was the first thing he and his father had done together in many years, and because they had been estranged for more than a decade, Rudolph wanted to preserve this important symbol of their reconciliation.

Rudolph and his father had been invited by the mayor to join the Committee of Fifty, and Rudolph took his briefcase and hurried the few blocks to the hall of justice on Montgomery Square. He had seen the membership list that had been put together by Mayor Schmitz, which represented the most prominent men in the city, as well as two women, Miss Katherine Felton and Mrs. John Merrill. Not included on that list were "the people's banker," A. P. Giannini; or the shadow mayor, Abe Ruef; nor any member of the board of supervisors. It seemed to everyone that the mayor wanted to run his own crisis management team, completely separate from the city's political powerhouses. Indeed, he was turning for advice to social and business leaders, and the only person who might be seen to represent the interests of the "common people" on the committee was the Catholic priest from St. Mary's Cathedral, Father Phillip O'Ryan.

When Rudolph arrived at the hall of justice, he found soldiers guarding the premises, but few citizens. He then saw a notice that said the meeting of the Committee of Fifty would be held at 3:00 in the afternoon. Rudolph returned to his office on Market and searched the street for his father.

He did not notice a cart filled with oranges passing him. Its driver was Amadeo Peter Giannini who, with an assistant, a cashier named Pedrini, was traveling toward his home in San Mateo, just south of the city. A second wagon, driven by a friend and bookkeeper, Frank Rossi, followed behind. An intelligent man, Giannini realized early in the course of the fire that he would have just one opportunity to save his new bank on the corner of Washington and Montgomery (now Columbus), and he quickly and unobtrusively loaded $80,000 in silver and gold coins into his wagon after sending his assistant to find as many oranges as he could bring back. Giannini, who was just thirty-six, had made a fortune in the produce business before turning to banking, so it was not a difficult task even in such great turmoil to acquire sufficient fruit and crates to cover the cash strewn across the back of his wagon. Transporting so much money through the chaos of an emergency—nearly $2 million in today's dollars—was a courageous act, but Giannini, who had never forgotten that his father had been shot dead in an argument over $2, also hid a pistol on the cart. He would at all costs protect the savings of so many hard-working Italians who had trusted their life savings to the Bank of Italy. The bank was only eighteen months old and, with $846,000 in deposits and loans, it was already considered a great success.

Giannini had learned that the government was closing all the city's banks due to the catastrophe, but he was certain that the Italians of North Beach were going to need his help to get through the tragedy. (He had no way of knowing that only three hundred buildings out of four thousand in the Italian section of North Beach would survive and that all of their owners were going to need money to rebuild.) He was well aware that he was developing the reputation of "a man who cared for the little fellows," and he had to be able to offer assistance if he was needed. On arriving at his home, he stashed his bank's ready cash in the ash container sitting next to his fireplace.

In a few days, even as the fires continued to burn, Giannini would construct a sort of teller's counter and bench near Washington Square and there in the open air would sign deals to rebuild homes and businesses in San Francisco when not another bank in the city had cash on hand. Their vaults were so hot they could not be opened for weeks, but Giannini was able to respond quickly and serve his people.

Historians would later portray the founder of the Bank of Italy as a victim of the cultural prejudice of the day against Italians, but Giannini was not the kind of man who would allow himself to be victimized by any kind of ill will or ostracism. He had married well, and when he left the fruit business he was, thanks to his realtor father's-in-law connections, named to the board of a local bank. He realized that the country was changing from a rural to an urban society, populated by newly arrived immigrant groups, that the big and established banks were not interested in small accounts, and that the "little fellows" would always need a trusted place to deposit their savings and to borrow against them. He knew his business and would increase it into the Transamerica Bank and then the Bank of America until by the Second World War it was the single largest privately held bank in the country, and no one was prouder than he that the average depositor held less than one thousand dollars in his account. Giannini had had the foresight to create a series of special departments in his banks—the Slavic department, the Russian department, the Portuguese department, and the Mexican department, for example—to meet the particular needs of various ethnic groups, and he loaned to them liberally to help them own their piece of a new country.

Rudolph Spreckels had a different view of his role. His was the world of wealth, privilege, and prestige, but one in which a larger responsibility to society was always taken for granted, the responsibility to serve the interests of honest business and an honest community—a true and sincere noblesse oblige. Not far from his Market Street office, he stopped on the corner where Montgomery meets Market Street, just in front of Hearst's Examiner Building, down from de Young's Chronicle Building—a corner that could be called the Times Square of the West. Many soldiers were patrolling the street, all with rifles over their shoulders or in their hands, the long, sharp steel of their bayonets gleaming for all to see. He looked down the street and saw that the Emporium, the largest store west of Chicago, was completely burned out, as was the Grant Building and Bancroft's History Building, along with many of his records and manuscripts. On the north side, the fires had already consumed most of the wholesale district, and all of its produce, dooming the city to shortages in all foodstuffs.

Suddenly, Rudolph was jolted by the sound of a large explosion. This was the first of many that would visit the city over the next three days, and Lieutenant Briggs, the army officer who set it, would later testify that he did not want to set off this explosion by using granular powder. He knew it was a dangerous procedure and that only dynamite would be safe, but he had been ordered by General Funston to bring down a building. This demolition at Kearney and Clay would shoot flame to the buildings across the street and cause yet another entire block to burn.

Where some smoke darkened the street because there was so little breeze, most of it rose in great columns to the sky. The fire was now being driven by its radiant heat northwest on Market Street, and the great temperature had broken a few of the lower floor windows of the Call Building. Rudolph watched as the flames entered the elevator shafts and rose past the Spreckels restaurant called the Rotisserie on the fifteenth floor and began to eat at the roof's windows just above the eighteenth floor. Soon, fire was shooting out of the flamboyant dome at the top of the building, and the tallest edifice west of the Mississippi began to burn completely.

Rudolph remembered how proud his father had been when, after buying the *San Francisco Call* in 1895, this building opened for the first time in 1896. The Call Building had been designed by the Reid Brothers in the manner of the Beaux Arts style of the American Renaissance, which shared the European horror of undecorated surfaces. Its ceiling beams were carved and from them hung exquisitely chiseled chandeliers, stretching from one end of the lobby to the other. Its pillars were Corinthian, and its walls sculpted. The furniture was of heavy English spool design and covered with velour and cut velvets. Rudolph, like Mayor Phelan, was a supporter of Daniel Burnham's City Beautiful movement, and the work of H. H. Richardson and Richard Morris Hunt. He was especially proud that his family had asked the Reids, who had designed the Hotel del Coronado in San Diego (the largest seaside hotel in the world), the Fairmont Hotel, and the Cliff House in San Francisco, to design the Call Building not to simply house a newspaper, but as a showplace for the city. Claus Spreckels's photo had appeared in all the newspapers at its opening and he beamed for the cameras. Its destruction would be a great loss for his father, for it had been constructed to be fireproof—steel beams protected by brick and mortar. If it was struck by fire, it had been predicted that the flames would be contained in one area until they could be extinguished, and so there was no need of comprehensive insurance.

Rudolph stood transfixed and was ordered several times by the soldiers to move along. But he simply shifted a few feet and continued his vigil there for more than an hour until the building was gutted from tower to lobby.

Rudolph noticed Abraham Ruef pass by in his high-backed automobile. The soldiers let him pass as if he were on a mission from the state house. He had his family in the car and was probably delivering them to safety at his yacht in the bay, Rudolph thought. Rudolph's brother John D. was at the same time taking his family down another street to deposit them in a suite on one of the great ships of his Oceanic Steamship Lines, a temporary residence that would provide them with water, food, and a help staff.

By now it was after noon, and Rudolph decided to take the briefcase to his home, to store the papers in his safe there. It never occurred to him that the fire could possibly travel all the way west to his neighborhood on the other side of Van Ness, and he did not yet know that the Hayes Valley fire was burning its swath across city hall at that very moment, heading toward Mission Dolores, at 16th and Dolores streets, the most important historical location in the city. It was here that two Franciscans said mass on June 29, 1776, which established the sixth of Father Junipero Serra's famous twenty-one California missions, and set the official birthday of the City of San Francisco. Mission Dolores was the city's center of Catholicism until the Secularization Act of 1833 transformed it into a private *ranchero*. Consistent with the city's colorful history, just after the Gold Rush days, the priests' quarters next to the mission became one of the bawdiest of the city's saloons. Fortunately, the fire skirted this priceless building, just four blocks from where it would finally be stopped three days later.

He reminded himself as he made his way to his home that in this great emergency he, and all San Franciscans, had to put aside personal commitments, enmities, and profit to work for the public good. Still, he could not remove Ruef from his mind, and any thought of the political boss of the Union Labor Party was bound to disconcert him. The audacity of Ruef was beyond that of any other man in San Francisco, a fact Rudolph knew from his own experience.

A $17 million bond issue had been scheduled for October 1904, and because the city had no debt to speak of it would have been easy for the city fathers to sell the issue at well above par. It was then that Ruef had asked to visit Rudolph in his offices, and when he arrived hat in hand Rudolph received him courteously until Ruef asked if he would be interested in buying the whole lot of bonds well below par. The political boss was trying to ensure against a boycott of the securities by an anti-union Wall Street and hoping that Spreckels would purchase the whole issue if the profit were guaranteed.

Incredulous, Rudolph asked how bonds being offered by a healthy city in a healthy economy could be bought below par?

Ruef explained that he would organize a strike of all the transportation workers the day before the issue and settle the strike the day after, a manipulation that would generate misgivings about the city's future stability, if only temporarily, and send the bond issue plummeting.

Rudolph immediately thought of the history of labor strikes in California and in the rest of the country, which, well before the Wagner Act of 1935 that gave workers the right to organize, were usually accompanied by violence. He responded to Ruef in as harsh a tone as was possible, "Do you mean to say that you would risk bloodshed for the mere sake of making money?"

Ruef then backed away and laughed. "No, no," he insisted, "I was only joking."

Spreckels remembered the countless number of times that men of business and status had complained to him in the sanctums of the Pacific Union and Burlingame clubs of being "held up by greedy politicians" and, having now experienced this venality firsthand, he sought out the head of the grand jury, T. P. Andrews. He asked Andrews how they could stop the corruption that was not only tarnishing the reputation of their city, but was also stymieing its business. Andrews replied that the matter would require proper investigation. It was important, Rudolph knew, to lay the first stone right in building anything, and to build a case against the mayor and his boss would require a careful mutual probing.

Rudolph had also taken his complaint to the editor of the San Francisco *Evening Bulletin,* Fremont Older, who had been railing in his editorials about the need for honesty in government. Older's wife had become concerned about a depression that she sensed rising in him, for he was thoroughly demoralized that his warnings had gone without any consequence at all. Rudolph, Phelan, and Cora Older all believed that a trip to the center of American power might bring the needed response and agreed that they and the city needed help in rooting out corruption. Older accordingly went to Washington for an audience with President Roosevelt.

The president had himself experienced many confrontations with corrupt political machines while a police commissioner in New York and so lent a sympathetic ear to the West's most important editor. In one way, Roosevelt must have been gratified to learn of the rampant abuses within the regime of the nation's first truly successful labor union party, a political hegemony he was very much against, and whose interests he was therefore predisposed to undermining. He told Older that he would lend the federal government's most experienced Secret Service investigator, William J. Burns (the future founder of the Burns Security Company), to look into the matter and that he would ask the famous San Francisco prosecutor Francis J. Heney to help, but that the costs would have to be underwritten.

No one was more prominent in the field of corruption clean-up than Burns and Heney, who had just returned from investigating the land schemes put forward by public officials in Oregon, which landed the governor of that state in jail. Older assured the president that the money could be raised, having in mind contributions from Spreckels and Phelan, and inquired as to what the costs might be.

When the president mentioned the figure of $100,000, however, about $2 million in current value, Older was less convinced that such a sum could be raised. A few days later, in trying to calculate the cost of the prosecution, Rudolph asked Heney what his fee would be. Heney replied that he was a San Franciscan and though he had no money to offer, he would donate all of his time and effort to the job. In the long run, and notwithstanding Heney's generosity, the costs would rise to more than $213,000, which was mostly contributed by Rudolph.

Ultimately, it was Abe Ruef who was at the bottom of all things corrupt in the Bay City. "See Abe" was the advice to businessmen seeking to do any kind of business with the city, whether to start a ferry line, to sell prison food, or to apply for the right to stretch telephone lines. Abe took the consultation, charging a "normal" legal consultation fee and retainer, usually fifty thousand dollars and often as much as two hundred thousand dollars. On occasion it could amount to a million. And, then, like feeding oats to the horses from a common basket, he divided the consultation among all eighteen members of the city's board of supervisors, always leaving a lion's share to be divided with the mayor.

Born of French Jewish parents in San Francisco in 1864, Abraham Ruef was a child prodigy who spoke American Sign Language as well as seven other languages, including Cantonese. He had a keen interest in philosophy and art and graduated from the classical studies curriculum of the University of California at Berkeley at just eighteen. Ironically, he was active while at college in the political reform movement of the times. He passed the bar and opened his own San Francisco law office in the same year he came of age to vote at twenty-one. Shortly thereafter he attended a meeting of a Republican Party club that was being formed in a shanty in the shadow of Telegraph Hill. Only two men were there and they told him that the meeting had just adjourned. Then they asked if he could write and if he would be the secretary of the club. It was a way into politics for a young, ambitious lawyer. Ruef accepted the secretaryship, writing and then delivering to the local newspapers an account of the meeting as it had been related to him. The account was published to the delight of the men, who then confessed that they had made it all up. Ruef laughed, realizing that he was finally in politics, real politics, and it was just a few years before he saw the opportunity of creating a political party to serve the interests of the growing labor union movement in San Francisco. Ruef's opportunity came in 1901 when James D. Phelan decided against running for a fourth term as mayor. When thirty-seven-year-old Eugene Schmitz, the head of the musicians' union and a violinist, was put forward to run, it was Abraham Ruef alone who was pulling the strings of the first labor union party in America to gain complete and absolute power in a city.

CHAPTER 46

By noon, according to the famed photographer Arnold Genthe, "the whole town was in flight. Thousands were moving towards the ferry hoping to get across the bay to Oakland or Alameda." The homeless who were not moving toward the wharfs were heading toward the open areas of Golden Gate Park. Genthe added, "The shock of the disaster had completely numbed our sensibilities."

Refugees were running to any piece of flat and open land they thought might give them protection—graveyards, railroad yards, anywhere away from buildings that might burn or fall. Some, like Jack Murray's family safely on the top of Telegraph Hill, looked out at the expanse of smoke before them being blown back from the sea, and did not feel threatened—not yet. Others carried suitcases on their shoulders or dragged trunks over the cobblestones. Union Square quickly became crowded with people choking in the smoke. If they had wagons they were piled high with furniture and personal effects. A Mr. Bacigalupi, who operated a record and music store on 4th and Mission, was hurrying back to check on the condition of his phonograph records when he passed one woman who was carrying her most prized possessions—her ironing board and iron. Another man was rolling an entire barrel of whiskey down Market Street. At his store Bacigalupi found a memorable incongruity in that the pianos in his showroom had collapsed on their legs while the rows of records in the next room were still filed neatly on the shelves. Writers who witnessed such events spoke of the stoicism of the people or of a mysterious insouciance that led them to accept their lot and prevented them from succumbing to misery. But it was not indifference that calmed them, or some particular San Franciscan inner strength. The fleeing hordes were traumatized to acquiescence by the sudden, terrible, and unalterable change in their lives.

At 1:00, James Phelan met General Funston for lunch at the Pacific Union Club, where Phelan was a member. The Pacific Union was then at Stockton and Post, at the corner of Union Square, and was then, as now, the city's most exclusive establishment. Its quiet, genteel, and luxuriant ambience must have been a startling contrast to the raging fires just to the north and south, not more than three or four blocks away and creeping slowly in their direction.

Since Phelan was a member it is certain that it was he who had asked Funston to lunch there. Though there is no record of what the two men spoke about, Funston undoubtedly wanted to convince the former mayor, who was also head of the Committee of Fifty, of the merits of his plan and tactics. Nothing in Phelan's notes or published writings seems to indicate any disagreement with the general's plan to have the city patrolled by armed soldiers and to dynamite buildings at will.

Phelan had much on his mind and many responsibilities to consider. He must have learned by then that the Call Building had burned and that his own building was in the path of the advancing Market Street fire, along with the Palace Hotel, the Chronicle, and the Examiner buildings. Phelan must have been thinking like a mayor as he listened to the general, concluding that the plan being put forward was from the hero of the Philippines, a war hero, and a recognized leader of men. The only thing that Phelan knew with absolute certainty was that the situation in San Francisco was becoming more desperate with every passing minute.

CHAPTER 47

A woman living on Hayes Street, just seventy-five feet from Gough, had arisen upon returning to sleep after the earthquake. She began to cook breakfast at about 10:30, not realizing that the shaking had caused her stove flue to shift, and the heat, instead of going to the chimney, went to her attic. That started what has come to be known as the ham and eggs fire, a benign and misleading title for a furious blaze that went quickly out of control because of the complete lack of water. From that location, called Hayes Valley, it burned east to Market Street and through city hall on its way to Mission. It would then burn southward over the course of the next three days from Ninth to Twentieth streets, where it would finally be stopped after burning more property in the city than any other. By 1:00, the St. Nicholas Hotel at Market and Hayes was burning down, as was the St. Ignatius church and college at Hayes and Van Ness.

There were now three distinct large fires burning in San Francisco: the South of Market fire, which extended from Sixth Street to the waterfront; the north of Market fire, which stretched from the Ferry Building to Sansome Street, traveling through the financial district, North Beach, Chinatown, and ultimately up through Nob Hill to Van Ness; and the Hayes Valley fire, which would claim city hall and the Mechanics Pavilion.

Earlier that morning, after the central emergency hospital in the basement of city hall had been wrecked by the quake, two doctors hurried across the street and forced the lock of the Mechanics Pavilion. It was a sizable space, the city's largest auditorium. Just hours earlier on this same expansive floor the young men and women of the city had been competing in a costume contest on roller skates, then a widespread fad in America. Now, the doctors determined, they would use the cathedral-size pavilion as a temporary hospital and had the surviving patients moved there from central. By 10:00, more than three hundred injured were lying, many on improvised stretchers of bare mattress or piles of blankets, from one end of the pavilion to the other and scores of the city's nurses and doctors had arrived to help in any possible way. Then, suddenly, the pavilion was swarming with thousands of desperate family members searching from one victim to the next, seeking the familiar face of a loved one. The dead were eventually moved to one part of the building while the injured were placed in another.

It would be just a few hours, though, before the Hayes Valley fire crept down Van Ness and leaped onto the roof of the Mechanics Pavilion. Every car, carriage, and wagon was then commandeered to carry the injured away, some to other hospitals, most to the Ferry Building, where ships had to be found to transport them to Oakland or Goat Island. The fire was threatening to enter the pavilion rapidly, and the nurses, all volunteers and many in civilian clothes, had to carry or pull each victim away in time to outrun the advancing flames. Finally, as the last group escaped the building, the fire breached it and soon filled every space. The reportage of this scene was much exaggerated in a story that was repeated by most major newspapers of the time. It first appeared in the *New York Herald,* and it quoted an Omaha man, O. K. Carr, who identified himself as a Red Cross worker who had witnessed nurses chloroforming patients and soldiers shooting them "as an act of humanity" as the fire overtook the pavilion. He claimed that 350 persons had been killed before the building burned down. Another supposed witness, Sam Cohn, said that several hundred persons had been burned alive in the pavilion, though he later admitted that the story was a third-hand account.

Some today continue to believe that the injured were tragically but necessarily left behind to die in rapidly encroaching flames. But two reporters, W. A. Mundell of the *Call* and Charles Brennan of the *Examiner*, refuted these accounts. Mundell, who later became a medical doctor, and Brennan, who later became an attorney, both worked as volunteers in the pavilion at the time of the fire, and both testified that they had together carried the last patient out of the building on a gurney. Mundell reported:

> . . . not a single injured person, not a single body, was left to perish or be burned when the Pavilion was abandoned about 11:30 in the forenoon of the earthquake. The evacuation of

> the building had commenced in ample time, and though the outer walls and roof were on fire in places and for brief periods while the evacuation was going on, no flames had penetrated the building itself until several minutes after the last of the patients were taken away.

Also, Captain H. H. Rutherford, a doctor from the Presidio, testified that he was inside the pavilion at 11:30 A.M. and all the patients had been removed. It was a day of extraordinary service at the Mechanics Pavilion and at other places in the city, but the newspapers delighted in the dramatic, and no story was too unlikely to be reported. Every excessive act seemed believable, given how incredible was the situation in which they found themselves. After all, one of the greatest cities in the world was burning down before everyone's eyes.

CHAPTER 48

At about 2:00 P.M., James D. Phelan's chauffeur, Jimmy Mountford, suddenly appeared in Phelan's motor car, explaining to his employer that he and the car had been pressed into ambulance service for the past few hours, carrying the dead and injured. They then drove to Pacific Street to pick up Rudolph Spreckels.

Ever the gentleman, Phelan inquired of Eleanor Spreckels, and Rudolph informed him that later in the afternoon, Eleanor, the children, and staff would go to a friend's house in Oakland. In another month, on May 26, 1906, Anna Claudine Spreckels would be born, a delivery that would be widely and erroneously reported in future history books as having occurred in a tent on the sidewalk fronting the Spreckels Pacific Street mansion during the days of the fire.

Phelan and Spreckels had planned to tour as much of the city as they could before the Committee of Fifty meeting at the hall of justice at 3:00 P.M. They discussed the policy of military occupation, which was already in force, and while both agreed that the pervasive presence of the soldiers was not a sound idea, they concluded that it would not be sensible to go directly against the mayor and General Funston in the midst of such turmoil. It would be better at the Committee of Fifty meeting to hold their counsel—on this subject anyway—and come together as a working group concerned only with relieving the crisis and not creating dissension.

As they were driving past St. Peter's Convent in the Mission district, Phelan and Spreckels came across a group of nuns, most of them older, who were evacuating their convent, carrying heavy crosses, statues, and other sacred items. They stopped the car, and Rudolph immediately ran to lift a large candelabrum from a struggling nun's hands. He took it to a waiting wagon, and proceeded into the convent to the salvage work and where he was soon joined by Phelan. They worked together with the nuns for more than half an hour, and as they labored they began to hear the continual booming of dynamite explosions. "Good God," Phelan said to Rudolph, "they'll blow the whole city apart if they go on like this."

Phelan left to pick up others he had arranged to chauffeur, and when he arrived for the appointed meeting at the basement of the hall of justice on Washington Square, he found the committeemen milling about. The meeting had been canceled again because of the fire, which by now was encroaching through North Beach. A new sign was posted, with instructions for the Committee of Fifty to meet later that evening at the Fairmont Hotel.

CHAPTER 49

By the evening of the April 18, the fire had spread through Portsmouth Square, and police officers worked feverishly to empty the hall of justice before it was taken. There was a morgue in the basement of the building, which quickly became overrun with corpses being delivered as they were dug out of the collapses. The police target practice range next door was then appropriated as a temporary morgue, but as the fire drew closer every sheet-covered body was removed, along with the vital records of the police department and piled high in the middle of the square. Canvas was placed over the bodies and records, and Detectives Charles Taylor and George McMahon were left to guard and preserve the valuable remains of humanity and history. The heat became unbearable, and embers continually lighted on the canvas and ignited it. McMahon broke into a bar on a nearby corner, taking as many bottles of beer as he could carry before the fire claimed the establishment as well. He used the beer throughout the period of greatest heat to cool the canvas each time it was brought to flames. When morning came it would all be saved.

Jack Murray and the men of Engine I worked the fire at Portsmouth Square but were forced to retreat, pushed back by the flames again and again, losing building after building. At the same time Lieutenant Freeman was directing his men—the sailors, marines, and firemen who were present at the scene on the waterfront—in pulling one hose line after another from the distribution valves of the navy's lone fireboat and several tugs. The lines extended up and down the harbor and down into the interior streets. Freeman had already organized a schedule of light boats to make continual round-trip runs from the harbor to Goat Island to bring freshwater to replenish the city steamer engines and to bring relief to a population that was becoming, in the pervasive smoke, mad with thirst.

The piers in San Francisco were, and are, numbered in twos from the Ferry Building, which stands at the beginning of Market Street. Even numbers go south and east, and odd numbers go west and north. Lieutenant Freeman made great progress in containing the fires around Pier 8 at Howard Street, and they were under control in this location by late afternoon. But Freeman knew that he could not rest for a minute, for blazes continued to rise all around him. A battalion chief with whom he had been working fell to the street in exhaustion, having worked the three-alarm fire the night before. Together, the chief and Freeman had saved the Sailors' Home, Folger's Warehouse, the Mutual Light Company, and many other business centers.

The heads of these companies would later write glowing letters of gratitude to navy and city officials about Lieutenant Freeman's extraordinary efforts, yet none of these testimonials seems to have been seen or endorsed by General Funston or Major General Greely, who would return to the city on April 22 to take charge.

In fact, later in April, a communication was received through the secretary of war from the White House with the following request:

> The President would like to have you obtain by wire at once from the commanding officers the instances of any special gallantry or signally efficient performance of duty by officers or enlisted men in connection with the San Francisco disaster. The President understands that two marines and possibly one or two artillerymen especially distinguished themselves in stopping the fire by means of gunnery or dynamite. The President would like those names at once.

General Funston had to have been asked for his recommendation, but in his view there were no awards of meritorious service that needed to be singled out. The closest the army came to acknowledging Freeman's efforts is a mention in General Funston's report, quoted in General Greely's report of the fire:

> The important work done by the Navy and the U.S. Revenue Marine Service in fighting the fire along the waterfront does not properly form a part of this report, as it was not done under my direction and control.

In the very next sentence, however, General Funston states that Admiral McCalla sent a force of marines "under Lieutenant Colonel Lincoln Karmany," which had rendered "excellent service independently on that day. . . ."

In fact, Freeman was definitely under the command of General Funston because Funston several times during the conflagration gave Freeman direct orders either to pick up dynamite or additional troops. The final word on the heroism of the day is included in Major General Greely's response to the White House request:

> With scarcely an exception officers and men spared neither personal exposure nor physical exertions in relieving earthquake and fire conditions. Fortunately the fire progressed slowly thus

> preventing great dangers through fire fighting. Dynamiting and blasting proceeded in rear of fire under unusual conditions indeed, but not involving great personal danger. Possibly my ideas of conspicuous gallantry and extraordinary services may be too high. In any opinion there were not such marked cases out of hundreds of officers and men who labored and fought strenuously as to justify me in naming these as superior to their comrades. General Funston fully concurs in this decision.

Major General Greely characterizes the fires as being slow-moving fire, which in his mind guaranteed the safety of those attending to them. Since he himself was not present he must have received this information from General Funston, though he does not realize that it is precisely because they advanced slowly that the army's order to evacuate the streets was a tactical error. Every able-bodied man, even before the full contingent of seventeen hundred soldiers arrived, should have been put into service saving the buildings floor by floor, emptying the building of contents to reduce the volume of fire, and standing at windows, two men to every window on Market Street, and down Van Ness, with wet mops and whatever water could be carried, in the time-honored tradition of bucket brigades. Men should have been stationed on every roof to stamp out the embers, or beat them out with brooms. In this way the fire could have been fought building by building, instead of relying on the San Francisco firefighters with their meager streams and uncertain water supply taking the initiative from the middle of the street. Just as there was a huge amount of wine that went unused, so was there a huge labor pool that was either unused or misused—men who were evacuated and reduced to onlookers, and soldiers with guns instead of work gloves.

It was a time for heroism, and not only the type of heroism that was acceptable to the army's sense of history. Every military man knows that gallantry is most often memorialized when it is associated with a single charismatic figure, like Leonidas at Thermopolae, Horatio at the Mulvian Bridge, Custer at the Little Bighorn, and Roosevelt at San Juan Hill. The days of April 1906 would go down as Funston's days, a testament of his command of the great earthquake and fire.

As General Greely further wrote in his report:

> General Funston and I are in accord in the belief that the conditions were not such as to offer opportunities for great personal bravery, or for especially conspicuous service. It is, however, my opinion that the conduct of General Funston and his command, almost without exception even to the last private, is deserving of the highest commendation.

As we have seen, however, history tells us that James O'Neil of Truck 1 had been killed doing his duty, as had police officer Max Fenner. Three firemen were pensioned because of the injuries they sustained during the fire: Fireman George Wells of Engine 12 lost his leg, George Woods of Engine 20 suffered a broken back saving the Hotaling Whiskey Company's warehouse, and Lieutenant Ed Lennon of Engine 2 sustained an inoperable rupture during the same collapse that tore Captain Brown's foot apart. At least eight other firemen were injured seriously enough to require treatment at various hospitals. E. Cosgrove of Truck 8 was in critical condition in the navy YMCA hospital, as was Gabrial Woods of Engine 20 at the naval yard hospital at Mare Island. Walter Cline of Engine 9 and Frank Shade of Engine 5 were also hospitalized. And at least one soldier, a Lieutenant Pulis, was blown up and severely injured setting dynamite. What more does it take to be recognized and cited as a hero?

CHAPTER 50

Lieutenant Freeman and his men now followed the fire south on East Street. They passed a great number of drunkards and miscreants who had congregated there, thinking perhaps that if the fire got too close they could always jump into the bay. They were obstreperous, and Lieutenant Freeman ordered a few of his men, sailors and marines, to move them at gunpoint.

By Wednesday evening the wind had suddenly begun to blow in stronger gusts across the bay, and Freeman saw that the fire was starting to spread rapidly over the foot of Rincon Hill and up its inclines. Although this section of San Francisco is now leveled, and above its flattened mass the high concrete approach to the Bay Bridge, early Bay City developers believed that this particular site had great potential for real estate development with beautiful, expansive views of the waterfront. The wealthy class had accordingly built there an exact replica of London's Berkeley Square, right down to the hedges and mews that they stocked with imported English sparrows. But time had worn it down since its heyday in the 1880s, and by 1906 the hill was overbuilt, covered with run-down flophouses for sailors, single-room occupancies for laborers, bars, and small factories.

To try to check the fire Freeman quickly ordered the *Active* to sail to the Santa Fe dock at Pier 26 at Spear Street. He sent the *Slocum* and the *Leslie* to the dock of the Pacific Mail Steamship Company at Pier 40, and to aid them he dispatched the revenue cutter *Golden Gate*. Hose lines were stretched from the boats. The mail company had a desalination plant, and the tugs were able to take on enough freshwater to feed the fire department steamers. They were also able to leave a two-hundred-gallon cask of water on the Mission Street Bridge for dehydrated citizens.

As he arrived, Freeman saw that the fire was running quickly up the sides of the hill and he ordered Midshipman Pond to have the residents evacuate immediately. Many, believing they were safe from the flames at that height, ignored the command. Some did leave, but reconsidered and returned for personal property. Eventually the flames completely encircled the hill and began traveling upward from every direction.

Midshipman Pond would later recall "the horrible situation, people ran screaming in all directions like a hive of ants whose hill had been disturbed. The old and the crippled, taken out of their houses on mattresses, were carried a little way and dropped, then picked up by someone else, carried a little farther, and then dropped again."

Freeman and his crew were overwhelmed by the ferocity of the fire before them. The heat grew ever more intense, and at every place they tried to break through with their hoses they were pushed back. Still, they rescued as many people as they could, dragging them through the smoke. In less than an hour the entire hill was in flames, and men, women, and children were trapped with no opportunity of escape. "The most heartrending sights," Freeman later wrote in respectful understatement, "were witnessed in this neighborhood, but with my handful of men we could not do as much for the helpless as we wished."

In this area of the waterfront, Freeman tried to enlist the help of those few along the waterfront who were sober, only to be rebuffed with a demand of forty cents an hour, paid upfront, a standard wage for someone earning $20 a week and not much different from the $1,200 yearly made by a fireman. Freeman wrote, "If I had had two hundred men at this time to aid in leading out hose and rescuing invalids and the aged, much more property and a great many more lives would have been saved."

The fire did not stop at Rincon Hill, and Freeman realized it was going to move on to the Southern Pacific Railroad sheds and terminal on Third and Townsend. He directed his men to abandon about 1,500 feet of hose line and then took the *Leslie* to the end of Fourth Street at China Basin. There he met some San Francisco firemen, and they provided Freeman with enough hose to stretch from the basin to Townsend, a distance of about three blocks. Freeman's men charged the line, and for the next two hours, with their lieutenant and commander yelling "Sock it to 'em," they held fast against the encroaching fire, armed only with the single line of hose being pumped by the *Leslie*. They fought like this until about midnight, when Freeman realized he had brought it under control. He and his men were utterly exhausted.

CHAPTER 51

Late on Wednesday, a telegram was received by the war department, sent by General Funston, pleading:

> We need thousands of tents and all rations that can be sent. Business portion of the city destroyed and about 100,000 people homeless. Fire still raging, troops all on duty assisting police. Loss of life probably 1000. [Note that the troops were not assisting in firefighting.]

At about that same time, Dr. Edward Thomas Devine received a telegram from President Roosevelt, asking the Columbia University professor of economics to be the chairman of the relief fund effort to aid San Francisco. Roosevelt did not want donations from America's people or federal aid money to be given over directly to the mayor and San Francisco politicians—surely not after his conversations with Fremont Older.

But Roosevelt also had more ambitious plans for the city's relief. Just a year earlier he had appointed future president William H. Taft, the then secretary of war, to take on the additional duties of president of the American Red Cross. Until then the Red Cross had been managed throughout the country by well-intentioned volunteers, but Roosevelt knew that it would be more effective if staffed with professional relief workers, managers, and fund-raisers. The U.S. Congress had just granted the organization its second national charter, and Roosevelt saw in the desperate situation in San Francisco an opportunity to show the nation how consequential the organization could be in peacetime. He therefore asked that Devine also go to San Francisco to represent the American Red Cross and issued a proclamation:

> To the people of the United States. In the face of so terrible and appalling a national calamity as that which has befallen San Francisco, the outpouring of the nation's aid should as far as possible be entrusted to the American National Red Cross, the national organization best fitted to undertake such relief work.

Devine was a wise choice because his book on the principles of relief was the standard text in the new sociological field of relief work, and the American Red Cross, founded twenty-five years earlier by Clara Barton, was the right organization to distribute the charitable funds and guide the relief efforts. Barton had recently retired, and President Roosevelt wanted to do all he could to ensure the continuation of her good work. As a start, Congress voted to allocate one million dollars to the Red Cross for its work in San Francisco. In all more than ten million dollars were contributed by the country's generosity.

As news of the conflagration spread east to the Atlantic coast, clothes and foodstuffs were also donated, which began to arrive the day after the quake. Even Mark Twain, the busiest of all public speakers, ended all of his appearances by saying, "I offer an appeal on behalf of that multitude, of that pathetic army of fathers, mothers, and children, sheltered and happy two days ago, now wandering hopeless, forelorn, and homeless, victims of immeasurable disaster. I say I beg of you in your heart . . . to remember San Francisco, the smitten city."

Not all the responses were positive, however. The stock market plunged on the news of the quake as investors sold off their California- and railroad-related stocks, and there were some indications of corruption—one midwestern state that sent eighty thousand barrels of flour to help feed the refugees learned that the barrels had been intercepted and auctioned off.

As so often happens, some of the relief funds also became subject to political machinations. James D. Phelan wrote of the Relief Committee of Portland, Oregon, which was withholding sixty thousand dollars that had been raised, refusing to send it in case some of the refugees from San Francisco decided to go to Portland, at which time the money would be given to them. The mayor of Portland was outraged and sent a public message. "If," he warned, "at ten o'clock tomorrow morning the money is not sent to San Francisco, you gentlemen had better flee to the country, because I will call on the mechanics in the shops and the farmers in the fields to come to the city hall with their implements and their tools, and we will manage to get it." The money was transmitted immediately.

Money would arrive from all over. European countries gave $300,000. John D. Rockefeller sent $100,000, as did his company, Standard Oil. Phelan raised $413,000 from 131 San Franciscans, all of whom were themselves in need of money to rebuild. The U.S. Congress sent $2.5 million. Phelan's finance committee ultimately took in more than $10 million in relief contributions, mostly through the Red Cross.

For the moment, though, San Francisco was in desperate need of supplies, despite the fact that the army's largest depot and supply warehouse west of the Mississippi was filled the morning of the eighteenth with goods of all kinds that, had appropriate action been taken, might have been used to relieve the crisis. In two buildings were stored thousands of sheets, beds, blankets, pillows, shirts, pants, shovels, and tinned food of all kinds, and though no horses were stabled there, there were dozens of carriages and flatbeds. Because of a peculiarity of army administration, however, the depot did not fall under the control of the Presidio or even under Generals Funston or Greely, but officially was the kingdom of Major Carroll Devol, who reported directly to Washington. Tall, dark-haired, pugnacious, and brilliant, Devol was the U.S. Army depot's chief quartermaster and supply officer. He would later reach the rank of major general and is considered by the keener historians of the American military to have been one of the best quartermaster officers in the early twentieth century.

As soon as the earthquake struck that morning, Devol rushed straight to the depot, at 36 New Montgomery Street, two blocks from his residence at the Occidental Hotel. In that neighborhood the quake's full fury had been unleashed. Wooden structures were a shambles, the walls or roof façades of some of the poorly built brick buildings had fallen out, and the first fires had immediately taken a foothold. Fortunately, the depot itself had survived without serious structural damage, and Major Devol immediately began rescuing what he could. Initially, he had little help, but gradually a few of the civilian employees and army clerks appeared to report for work. The army's second warehouse, the commissary depot, a few blocks away, had already gone up in flames.

The first clerk to arrive, O. D. Miller, found the rear of the depot on fire. He and Devol, aided by a few workers, salvaged as many records and files as possible and deposited them in a wagon parked outside. Devol then directed his attention to the safe but, unfortunately, its combination could not be found. No one but the chief clerk, Isaac Onyon, knew it, and Miller was sent running to Onyon's home on Ellis Street, nearly a half mile away. Upon arriving, completely out of breath, Miller found that Onyon was not there. His wife advised that he would probably be found at a nearby establishment. There Miller discovered Onyon standing at the bar, calmly drinking a glass of Anchor Steam Beer. After an exchange of pleasantries, Miller asked for the combination, which he wrote on the back of an envelope. It was an eerie, businesslike transaction in the middle of a rising chaos. Miller then ran back to the depot as Onyon continued his drinking until a few hours later his residence burned down.

While Miller was retrieving the combination, Devol dictated a telegram for the general quartermaster of the war department:

> Terrible Earthquake at 5:15 this morning, buildings on fire all over lower part of City, no water at Mission St., 2nd Comsy

> [Commissary] Depot burned to the ground. Office building—Storehouse at 36 New Mont. St. now on fire, small hope of saving.

As almost all telegraph communications were down, Devol sent the message by boat to Oakland, where it was to be transmitted to Washington, D.C. Soon afterward the troops from the Presidio arrived at the depot, probably just after 9:00 A.M.

When Miller returned to New Montgomery Street with the safe's combination he was refused entry by an overzealous guard, and together they had to wait for a superior officer. As he recalled, "Troops under command of General Funston were in charge and would not allow me to cross the street. After identifying myself and explaining to a Captain in command of the guard—which was engaged in preparations for dynamiting buildings—he passed me across the street. Major Devol and I then entered the burning building and after groping thru smoke, reached the office safe. He read the numbers off to me as with trembling hands I worked the dial—the door fortunately opened at the first attempt. I then took the cash which was in a tin box, and the official check-book and cash-book."

Just as the two men exited safely, the building became totally engulfed in flames. They rushed back to the major's residence at the Occidental Hotel, where Devol had sent the single wagonload of rescued items. Over the next several days the army was forced to commandeer many wagons, having lost so many of their own transport vehicles at the depot.

CHAPTER 52

As the day wore on, General Funston worked diligently, moving in his high-backed automobile, a hired Pope-Toledo, from one location to another, all the while determined to remedy the deteriorating situation. The car with its driver had been rented for one hundred dollars a day (about two thousand dollars a day in current dollars) from a motoring enthusiast,

William Levy of the Levy Bag Company. It was a formidable car, fit for a general, and price gouging seemed to be acceptable, at least in this case. A sign announcing U.S. was placed on each of its back doors to identify the vehicle and prevent its confiscation by army officers in need of transportation. The commandeering of automobiles was occurring regularly in the city, and on that first day, there was a general and reasonable, if illegal, order to impound every car.

The forces at Funston's immediate disposal at this time were probably less than six hundred men, most from either the Presidio's artillery corps or Fort Mason's engineering corps. Funston began calling in troops from the other coastal batteries, such as Fort McDowell (just across the Golden Gate) and Fort Miley (just south of the Presidio), but most of the veteran infantry and cavalry troops had been dispatched elsewhere. The two largest contingents were the 22nd Infantry Regiment, which was stationed at nearby Angel Island, and the 20th Infantry Regiment (not to be confused with Funston's 20th Kansas Volunteers), stationed at the Presidio in Monterey, approximately seventy miles to the south. Both the 20th and 22nd had arrived in California only a few weeks earlier, each having returned from the brutal and bloody fighting in the Philippines.

Most of the troops of the 22nd were encamped across the bay at the army's rifle range near Point Bolinas in Marin County, and Funston sent a large tug, *Slocum,* to begin ferrying them to the city. But it was more difficult to communicate with the Monterey soldiers since the telegraph system was down. Funston directed Captain Leonard Wildman, his ranking signal corps officer, to try to restore communications with the outside world.

Anxious to bring the several hundred men from Monterey to San Francisco, Funston needed a large, fast vessel and decided on the *Preble*. It is uncertain if the navy provided or Funston commandeered the *Preble,* but in any case it was at this time that Lieutenant Freeman decided to remain on the wharves to fight the fire and assigned a warrant officer to pilot the ship under full steam to Monterey.

At around 10:30 A.M., Funston remembered or was reminded that the offices of the Commercial Pacific Cable Company were in the Grant Building, next door to the administrative offices of the army's Pacific Division. Unlike John Mackay's Commercial Cable Company, a private company that had laid cable beneath the Atlantic, the Pacific Cable had been created by the U.S. government and then leased to a private contractor. The cable had not been damaged in the earthquake, and people not put off by the horrific expense of sending messages to anxious relatives or business contacts by transmitting them first around the world were using, or trying to use, it that morning. The chief clerk, however, could not calculate the proper charges for the unusual requests, having never before sent messages by this highly circuitous method, and so they were sent without fee.

Funston sent his troops to impound the offices of Pacific Cable, and private communications in San Francisco were halted at about 11:00 A.M. The clerks and telegraphers would now be working for the army. It was at this time that Funston sent a telegram to Secretary of War William H. Taft, informing him of the disaster. It took awhile before Taft was able to read the message as it was first sent to Hawaii, routed to the Philippines, transmitted to Japan, then to Russia, through Europe, and on to Washington. Much international intrigue and spying was taking place during this period, and governments of most of the world's powers were able to read Funston's dispatch before the officials in Washington.

Taft drafted a response ordering Funston to communicate only by cipher, which created some confusion, for the cipher books were destroyed when the fire reached the Grant Building later that day. Consequently, when Washington sent messages in code to Funston, they arrived as meaningless groups of five digits, each separated by a double space. General Funston finally had to ask the war department to transmit its dispatches unencrypted.

CHAPTER 53

By the time Funston returned to the Phelan Building after his lunch with James Phelan, troops of the 22nd Infantry Regiment had begun arriving in the city from Angel Island and Marin County. The 22nd was one of

America's oldest fighting units, having been formed in the War of 1812 and served with distinction in the Civil War. However, its service record after 1865 deserves some scrutiny. It had been heavily involved in suppressing the Sioux uprisings in 1876–77 and 1890, which is to say the Battle of Little Bighorn and the Battle of Wounded Knee. When the famed 1892 Coeur d'Alene mine strike in Idaho got out of hand, it was the 22nd that was sent in to quell the disorders, participating in the only instance of declared martial law since Lincoln's illegal imposition of it during the Civil War. In the Coeur d'Alene operation, the 22nd arrested numerous civilians, not just the strike leaders, but hundreds of sympathizers as well, working with lists provided by the mine owners and local police. A few years later, the 22nd would fight in the bloody guerrilla conflict in the Philippines. Both the Sioux uprising and the Philippine insurrection were remarkable for the high number of civilian casualties involved—three hundred at Wounded Knee alone and more than two hundred thousand Filipinos killed or disabled by the war. The 22nd, therefore, came to San Francisco with a history of involvment in dubious occupation situations, in war and in peace.

No sooner had troops entered the city than reports began circulating of instant executions. The mayor's infamous shoot-to-kill proclamation had been sent to the printer early in the morning, but it wasn't until noon that the five thousand handbills began to be handed out, tied onto streetlamps and nailed to the doors of buildings. However, soldiers knew of the order as soon as it was made by the mayor, and it was reported earlier that, acting under General Funston's orders, they had executed "about a dozen looters." Funston later admitted to knowing of two killings for looting, but that he had never investigated them.

"Through all this terrible disaster," he wrote in his report, "the conduct of the people had been admirable. There was very little panic and no serious disorder. San Francisco had its class of people, no doubt, who would have taken advantage of any opportunity to plunder the banks and rich jewelry and other stores of the city, but the presence of the square-jawed silent men with magazine rifles, fixed bayonets, and with belts full of cartridges restrained them. There was no necessity for the regular troops to shoot anybody and there is no well-authenticated case of a single person having been killed by regular troops.

"Two men," he concludes, "were shot by the state troops under circumstances with which I am not familiar, and so I am not able to express an opinion, and one prominent citizen was ruthlessly slain by self-constituted vigilantes."

In the three days just after the earthquake, while the fire raged, no military or civic official denied that lawbreakers were being shot in the city. Mayor Schmitz told reporters on the first day of the disaster that to his knowledge three looters had already been killed. It was only after the catastrophe, when the legal consequences of officially sanctioned murder began to be considered, that officials did their best to cover up the facts. Reports filed by journalists in the first several days of the disaster were denounced as exaggerations or outright lies.

Many formal investigations of the actions of the military were made after the fire. Insurance investigators alone interrogated hundreds under oath. A manager of one of the city's largest stores, A. J. Neve, said the "looters caught at this work were shot without question." A photographer who worked for the *Chronicle,* Moshe Cohen, said that he saw five bodies "shot on the brick; they'd been shot and left to die there." A miner, Oliver Posey, testified in a deposition that "instant death to scores was the fate for vandalism; the soldiers executed summary justice." Like all statistical evidence for the four days of fire, the number of civilians executed by soldiers varies widely from account to account. Funston himself admits in his narrative to three, and there are accounts that claim as many as five hundred were killed. Based on eyewitness testimonies collected with extraordinary care by the director of the San Francisco virtual museum, Gladys Hansen, the total number of people shot down by the military is estimated to be at least five hundred—one sixth of the three thousand victims the virtual museum cites as having perished in the earthquake and fire. Any educated analysis of the period would suggest that the actual tally is much closer to Ms. Hansen's figure than to General Funston's. Any analysis, however, has to take into account the fact that this was a tragedy witnessed by many writers and photographers, and if as many as five hundred people were shot by soldiers it can be assumed that there would be many more firsthand accounts of those shootings than actually exist. Of course, the treasure trove of historical papers that were gathered by the Committee of Fifty, should they ever appear, will certainly tell a fuller story.

CHAPTER 54

That martial law had been declared was understood by most San Franciscans, some of whom viewed it as license to take the law into their own hands. One of the notable casualties it provoked involved a prominent citizen by the name of Heber Tilden. He was the first to arrive at the Red Cross after the earthquake, offering himself and his automobile in assistance, and he transported victims from hospital to hospital, with very little rest, through the days of the catastrophe. On the Sunday following the earthquake Tilden had been ordered to stop his vehicle, which had the large flag of the Red Cross flying on it, by a member of a citizens' patrol on 22nd Street and Guerrero. Tilden kept on, thinking the man would certainly see the insignia waving from the car's fender. By then the streets had been converting slowly to some orderliness, and Tilden must have relaxed his guard. The patrolman and two others standing with him took it on their own to shoot at the passing car, and one of the shots killed Tilden instantly. The three were arrested and found to be members of a group called the Citizens' Police, men organized to help the police without being given any particular authority, let alone the authority to shoot people.

At their trial, Superior Judge Carroll Cool charged the jury by explaining the conditions of the most recent social history of San Francisco. As a matter of law, he informed them, at the time of the emergency "martial law did not prevail" and "mere proclamations could not make law."

But he also added,

> It is a matter of history that the entire community believed that martial law prevailed during the great fire. Therefore, if the defendants honestly believed and the circumstances were such as to lead them to believe that they were acting under martial law, and the evidence proves that that mistake removes

any criminal intent, then the defendants were incapable of committing this alleged crime.

The jury deliberated just a few minutes before finding that the Citizens' Police acted in good faith by killing the ill-fated Red Cross volunteer.

The reported killings continued until after the fire was finally stopped three days later. It has been determined that most civilians shot on the streets were killed by soldiers of either the U.S. Army or the California National Guard, and since both wore similar uniforms, it was difficult for the witnesses to know which service was involved—a fact that stymied subsequent investigation.

Although four different services were involved in those four terrible days—the U.S. Army, the U.S. Navy, the U.S. Marines, and the California National Guard—the tone was set early in the disaster by the army. The troops forced the evacuation of thousands of home owners and shopkeepers, those most motivated to risk their lives to save their property. By the second day, the national guard had begun to see the ineffectiveness of such policies and started to enlist the help of local residents to fight the fire. But the army continued in following the evacuation order to the last. The navy, under Freeman, welcomed volunteers on the waterfront and in the North Beach district, a policy that helped contain the fires in those places.

Much has been written about the cultural differences between the army and navy. Fundamentally, the navy spent its wars on the seas, distanced from its enemies, and the enemy thus became a more abstract target than human life. The army, in contrast, often found itself eye to eye with other human beings who meant to do harm, and its distrust was made acute by such experiences. This distrust was brought from the fields of the Indian Wars and the jungles of the Philippines to San Francisco by soldiers who saw the world in very different terms than we do today.

People were also killed by marines, and by "special policemen," volunteers who had been serving the city since the 1850s. "Special police officer" has a specific meaning to the SF Police Department. Even today "special police officers" wear uniforms and patches that are virtually indistinguishable—to all but careful inspection—from those of standard police issue. The regular police wore a seven-point star on their jacket breast, but a special police officer wore a six-point star that had two or three letters in the center rather than a number. The first time a particular badge was assigned, the letters represented the wearer's initials, but when it was returned, it was usually reissued to someone whose initials did not match.

A Berkeley collector, Michael McDowell, owns a letter written on a cut brown paper bag and sent without an envelope through the mail in April 1906. It is from Mary Doyle of 85 Henry Street and was received by her cousin in Chicago. "Excuse paper," it begins, "nothing else to write on. . . . A large number of men and even women have been shot down for disobeying orders of soldiers. . . . We are camping in an empty flat with nothing but mattresses on the floor, cooking on the street, no fire allowed in house. . . . Lines of people three blocks long waiting for bread. . . . Water is all putrid. . . . City is an entire waste, everything ruined or burned. I consider we are very lucky to escape with our lives. . . . Had to get out fifteen minutes after the shock. No time for anything."

Private De Loa Kaff, in a letter to his family, wrote, "Looters are being shot left and right." Another soldier wrote home that "strict martial law prevails and the soldiers run everything and every night a number of people are shot by the soldiers . . . people caught looting are shot dead without warning."

However, he assures his mother that "there is not much danger of me being shot because I do not give the people I halt any chance to get the drop on me. At night you'd think there was a war going because shots are being fired all the time."

Major Carroll Devol's daughter, in a letter sent to a friend in the Philippines, wrote, "The soldiers have shot any number of men who were seen thieving. . . . The thieves are at work and the soldiers shoot them on sight. . . . A good many awful men are loose in the city, but the soldiers shoot everyone disobeying in the slightest, no explanations asked or given."

Francis Whittaker tells of one incident of a grocer who violated the mayor's injunction against selling bread at above-market value. An army officer seized the bread and began giving it away to the needy. The retailer protested that these goods were his own property and ordered the soldiers to leave. The officer ordered his men to take the grocer outside, where he was then shot against the wall of his own store.

Andrea Sbarbono, the president of the Italian Swiss Colony wine company, saw an Italian immigrant killed in an altercation with a marine. Since the incident took place in a section of the waterfront that had been spared from the conflagration—where fire would not later cover up the evidence—the soldiers wrapped the body of the slain man in chains and dumped it in the bay.

Mr. Guion H. Dewey, a Virginia businessman who was staying at the Grand Hotel, reported that he "saw innocent men shot down by the irresponsible militia." He also described seeing "a soldier shoot a horse because its driver allowed it to drink at a fire hose that had burst."

The photographer Moshe Cohen also saw a man with his arms full of goods take fright at the sight of three soldiers and start to run. The man was immediately shot down in cold blood. "How did they know that he wasn't entitled to the stuff?" wrote Cohen as he later recounted the events that day. "He could have worked in the place, or maybe even been one of the owners. Maybe he ran because he was scared and as sure as hell he was entitled to be with three guns pointing at him."

The casual manner in which the army and national guard killed people sickened even the members of the other services. One marine from the USS *Marblehead,* Daniel Morgan, gave a long account to a reporter about the "raw" killings he saw soldiers commit.

Later in the week, even the army's own local newspaper, the *Presidio Weekly Clarion,* published an account of a shooting:

BANK CASHIER SHOT TRYING TO OPEN SAFE WITHOUT PERMISSION

> The cashier of a Market Street bank was killed Friday morning, April 20, by a soldier. The cashier was engaged in opening a vault. Military regulations require that application be made to General Funston, who directs the officer commanding the ruins to send a detail of soldiers to the safe sought to be opened. Unless these formalities are observed no safe can be touched.

No mention is made in this article of whether the man was engaged in theft or simply following the instructions of a superior. The same event was covered in the *San Francisco Bulletin,* where it was reported that the soldier fired without any warning.

The first of dozens of books written about the earthquake and fire (and perhaps the first fast-tracked book in publishing history), *San Francisco's Great Disaster* by Sydney Tyler, published just a few months after the catastrophe, reported on other victims of martial law: "General Funston realized that stern measures were necessary, and gave orders that looters were to be shot at sight. Four men were summarily executed within six hours."

In one incident an Elmer Enewold called to a man he saw a half block away, picking over the rubble of a ruin. The man did not respond, so Enewold fired a warning shot, at which point the man began to move away from the scene. At the same time, however, a regular soldier heard the shot, saw the man running, and concluded that a crime had been committed and shot the man through the neck. In fact, the victim had been working his way through the stones to reach a man who was half buried, perhaps even half alive, in the rubble.

A medical doctor reported watching as a drayman demanded twenty-five dollars for hauling a family's possessions away while others had been charging five dollars. The property was already loaded on the carriage bed, and he refused to take less. He also refused to give up the haulage. Soldiers appeared and ordered the man either to return the possessions or to haul them for five dollars. When the drayman declined to do either, a soldier ran him through with a bayonet. Another man was reported to have been killed by a reservist who ran a bayonet through him for rummaging in the ruins of a jewelry store on Market Street.

It wasn't long before stories of official murder and mayhem spread though the city, magnified and adorned with each retelling. Tales were told of ears cut off for the earrings, fingers for the rings, people being shot randomly at windows to save them from the fire. Fourteen people, it was rumored, were shot while robbing the U.S. mint.

San Francisco may have had a few rough edges, but the heart of the city was primed and pumped by good, fair, and hard-working people. The mythology of wanton murder and widespread criminality grew from a general desperation within the event itself, and fed by a national frenzy for news, the more sensational, the better.

Situational awareness is a term used by first responders and the military to describe a set of circumstances that is not perceived equally by people who experience it. It can cause different, even contradictory, behaviors in an emergency or battle environment. In the San Francisco catastrophe people generally believed they were in a legitimate shoot-to-kill situation, and that where looting was concerned there was every reason to believe the worst about their neighbors. General Funston, an honest, forthright, and upstanding individual had helped bring about such a "situational awareness" among San Franciscans by making short-sighted decisions.

CHAPTER 55

Shortly after the fire, Henry Anderson Laffler, an assistant editor of *The Argonaut,* a highly regarded literary magazine, wrote an article challenging the account of the crisis that General Funston had published in *Cosmopolitan.* His managing editor, however, refused to run it, for *The Argonaut* wanted upbeat pieces about a bold new city that would arise from the ashes of the old. Laffler then published the article in booklet form, and all of the incidents he cites are confirmed by other eyewitness accounts.

In his essay, Laffler comments on the map published after the disaster that shows not only the areas in San Francisco that had burned after the earthquake but also, in white, the areas that had been saved. "One looking at this map," he wrote, "is instantly impressed by curious dots, flecks and irregular lines of white scattered over the otherwise unbroken red area that stretches from Market Street north to the waters of the Bay, from Van Ness Avenue eastward to the docks and wharves; and one is irresistibly impelled to inquire by what strange chance, by what seeming miracle, these detached and isolated structures still stand unscathed amid the surrounding desolation?"

Laffler then goes on to explain that most of these islands of white dots and flecks represented stands made by local citizens against the fire. One was a house owned by O. D. Baldwin, a man thought "no longer young in years . . . determined to save his home if it could be saved."

> On the morning of the first day, Mrs. Baldwin had instructed her servants to fill with water all available vessels—even the porcelain vases in the drawing room—and these had been distributed throughout the house. This water now served, in the hands of Mr. Baldwin, his son, his nephew, and the Japanese servants, to extinguish the thin tongues of flame which appeared each moment on woodwork fronting the fiercely burning residence in the same lot not twenty feet distant. A hole was broken through the slate roof. The blazing cornices were being chopped away—when appeared a military officer who ordered Mr. Baldwin and his helpers instantly to leave the house! But Mr. Baldwin did not leave the house. A refusal was on his lips when the officer recognized in Baldwin an old friend, a one-time neighbor, and, on the instant, the soldier carrying into execution the unlawful orders of the military authorities became a man eager to help his friend.
>
> "Mr. Baldwin! What can I do for you?"
>
> "Drive out this rabble," said Baldwin, "and help me."
>
> At the point of a pistol, the soldier drove from the house the men (looters, shall we say?) who had poured into it, and in a short space of time the fire was extinguished, the house saved. Every window in the building was broken by the heat; furniture which stood in the center of large rooms shows blisters on its polished surface. Yet the house stands, a monument to the courage of its owner and defender.

Laffler then went on to describe how a number of houses on Green Street, not far from the Baldwin residence, were saved:

> . . . the fire which was moving steadily from south to north along Leavenworth Street and along Jones Street attacked the southernmost line of houses in the block in question. The residents of the houses upon the north line of the block, perceiving that the flames were advancing slowly, determined at least to make a fight for their homes. Prominent among these men was Dr. J. K. Plincz, a young surgeon, and Mr. Kirk Harris, formerly of the staff of the San Francisco *Chronicle*. These

> two men with a few others, some passers-by, and two or three carmen from the Union Street Car-house, set to work. They chopped and broke down fences and small outbuildings that might afford a pathway to the fire; they achieved the successful destruction by dynamite of a small barn; they wet blankets, rugs, and carpets with small quantities of water that had previously been collected in pails and bath-tubs, and one by one, as sparks fell or shingles caught, they beat out the flames. A dipper of water here, a stroke with a wet cloth there—that was all—enough.

Laffler then recounts how this triumph was almost lost:

> But soldiers are also men. Not infrequently they hesitated at carrying out the unlawful orders which went far to work the destruction of our city. And so it was here. Dr. Plincz, laboring to save that fine old octagonal house that stands at 1027 Green Street was ordered by a soldier to leave it. A little persuasion, a diplomatically assumed air of camaraderie, a few glasses of good wine—and the soldier was ready to countermand his order and go away. Had the soldier been a little more stubborn, had Dr. Plincz been a trifle less diplomatic, what happened a thousand times in San Francisco during these terrible days would have happened here: the defenders of these homes would have been compelled at the point of a rifle to abandon them to a fiery fate. But the soldiery was not the only human peril with which these men had to contend. Appeared on the scene a youth, wearing a badge of authority, who casually said to Harris of the *Chronicle*:
>
> "Well, in a few minutes these houses will be up in the air."
>
> "What!" said Harris.
>
> "Yes," responded the youth. "We have decided to dynamite this block."
>
> "But surely," said Harris, "you fellows are trying to save property with your dynamite. I know that you are not trying to destroy property. And these houses, as you see, are already out of danger, and are not dangerous to any houses anywhere."

"Well, we've decided to dynamite them," remarked the youth, "and they are going to be dynamited."

And then it was that Harris hung about the man's neck and brought his neighbors to labor with the youth, and after long minutes of argument and appeal won the boy's grudging consent not to raze utterly these homes that had so heroically been saved from fire. With infinite relief they watched his departure to sate elsewhere his lust of destruction.

If diplomacy did not work, evasion could sometimes have the same effect:

Another blotch of white upon the map of red represents the houses that stand on Russian Hill. Here are the residences of Mr. Stone, of Mr. Morgan Shepard, of Mr. Richardson, of the Rev. Joseph Worcester, of Mr. Livingston Jenks, of Eli Shepard, the three-story flats the property of Mrs. Polk, the property of Mrs. Virgil Williams, the house of Mr. Livermore, and several other homes scattered over three city blocks, with three or four houses a block away to the north.

The fact that these houses stand is owing only in a slight degree to their situation. It is owing not at all to any fireman. It is, for the most part, in despite of the military. A score of times these buildings were on fire. A score of times the men of the hill, evading or resisting the efforts of the military to expel them from their homes, extinguished the flames. Thursday night, not far from eight o'clock, the soldiers swarmed upon the hill. They ordered everyone to depart. Mr. Norman Livermore, who had collected for the defense of the Livermore residence perhaps a barrel of water, eluded the man with the gun who went tramping through his residence, and, when he had gone returned to it. An hour later a row of wooden houses on the crest of the hill, separated from his residence only by a narrow alley, were burning fiercely to its constant peril. Had he been absent a heap of ashes would mark the spot where the house now stands. And once again, before day dawned, Mr. Livermore was ordered to desist from remaining in the vicinity

> of his house, and was compelled to return to it, without knowledge of the military, by devious ways.
>
> It is to be feared that the readers of these lines will weary of the monotony of this narrative. . . . It might be narrated in detail how Livingston Jenks was driven at the point of a pistol from the roof of his mansion where he had entrenched himself; it might be narrated at length how the house of Mrs. Morgan Shepard, through that long, intolerable night, was invaded by soldier after soldier who commanded her to depart, and whom by feminine tact and diplomacy she managed to restrain from violence, and thus remained beneath the shelter of her roof throughout the fire; it might be told with circumstance how Mr. Richardson and his wife were forced at the rifle's point to leave the hill, abandoning their homes to the flames, and returning to find it scorched indeed, but saved from destruction through the efforts of their neighboring friends.

The question can rightly be asked, if a few citizens could accomplish so much, what would have happened had entire neighborhoods been allowed to remain in their homes? Even when San Francisco firemen tried to recruit civilian volunteers, the army often attempted to prevent the recruited men from entering the lines. To the soldiers, once an area was evacuated, any civilian reentry was prohibited.

Accounts of looting most frequently involve areas that were about to be destroyed in the fire and that had therefore been evacuated by the military. Most of what was looted was about to be burned anyway—an observation noted often by members of the military. Since the army troops were instructed to keep people at least two blocks away from the fire, large tracts of land were vacated. It was that property that was reported in many instances to have been looted by its protectors, the army and national guard.

Laffler recounts such episodes:

> From the Durbrow house which still stands at the corner of Leavenworth and Francisco Streets the soldiers drove [out] its defenders, saying that the house was to be destroyed by dynamite, and when, hours later, they returned, they found a man in the uniform of a United States marine who had ransacked

> the place. He had taken money from a purse that in the haste of departure had been forgotten. But the soldiers refused to arrest him; he was permitted to escape. Next door lives a Mr. Marples, who also was ordered from his home, and who also, when he returned to it, found a looting soldier—a looting soldier, in fact, with the hardihood to depart with his bag of loot despite the presence and entreaties of its lawful owner. It is a common story. From the adjacent residence (at 828 Francisco Street) of Mr. A. C. Kains, of the Canadian Bank of Commerce, the soldiers endeavored to drive [away] all persons at noon on Friday, but in one way or another the Kains managed to remain till late that afternoon. They locked the doors, they closed the windows, but nonetheless, a soldier was throwing loot from the window of the drawing room when Mrs. Kains returned. "This house is abandoned," said the soldier in response to her expostulation—the [very] soldier who had ordered them forth! The stories of the ordeals through which passed various other houses of this section—the Fontana, the Hume, the Fay, the Copeland and others do not essentially differ from those already related. They are one and all but variations of the same tale.

Many other similar incidents were reported. San Francisco police officer J. C. Schmitt watched dumbfounded as troops looted his home. Liquor establishments that had been closed by one group of soldiers would be pilfered by others. Midshipman Pond, working with Frederick Freeman's group of sailors and firemen trying to save North Beach, discovered two drunken soldiers trying to rob a jewelry store and was almost shot by one of them as he and another seaman tried to apprehend them.

Though Laffler cites specific events that were attested to by a number of respected citizens, Funston would later dismiss his essay as a "cowardly" attack full of "mental gymnastics and fairy tales." However, a number of merchants whose businesses had been looted by the troops were able to gather signed affidavits from a number of people, including firemen and policemen who witnessed the thefts and used these affidavits afterward to sue the federal government for damages.

Also, the unrestrained use of dynamite caused equal consternation and criticism among San Franciscans. One of the city's leading citizens,

Osgood Putnam, is quoted as saying, "Untold grief would have been spared, and I believe in many cases, mental imbalance has resulted from keen personal losses. Hundreds of cases can be duplicated where the senseless [dynamiting] was carried out, keeping people from their homes for many hours before the fire reached them."

Another citizen wrote, ". . . the soldiery had driven all men who might desire to fight bravely to the last in their homes' defense . . . [which left] streets whose awful silence was even as a cry of agony before impending doom. . . ."

In his early account of the fire, Sydney Tyler wrote, "True to his record, [General Funston] once more 'turned the trick' and Mayor Schmitz only expressed the opinion of 450,000 other residents of the city when he said: 'The Army and the nation are to be congratulated on the possession of such officers as General Funston.' "

It was an observation that would not stand the test of time, at least in regard to the fire.

CHAPTER 56

At about 3:00 P.M., Captain Wildman finally established communication with the outside world at the telegraph office located in the Ferry Building. Information was secure now and could be transmitted. But the fire continued to consume a more precious form of information—the city's cultural legacy. All of San Francisco's public library buildings were lost—the main branch, the Phelan branch, the North Beach and Rialto branches. In all, 166,344 books were lost.

At about this time, it became evident that the dynamiting was not going well. The blaster Bermingham, who was so often alleged to have been drunk, would later testify in court that he saw sixty buildings blown up on the first day alone, and in each case, a fresh fire was started. It is difficult to find more than two accounts of demolition by dynamite that were successful, yet there are many reports of fires that were started by the dynamiting policy. It is telling that Funston remained determined to use explosives to stop the conflagration, even though plenty of evidence argued against that strategy by the early afternoon of the first day.

With the fire making its way up Market Street, Funston now ordered the evacuation of the Grant Building. The instruments for the Pacific Cable Company were removed from there and transferred to an equipment shed at the western end of the city. There, the telegraphers were able to tap into the cable and make a fresh connection, and communication with Washington was reestablished. Pacific Cable continued to be functional throughout the disaster.

Earlier, as it became evident that the fire's progress would not be stopped, Funston relocated his headquarters from the Phelan Building to Fort Mason, where he would remain for the most part throughout the conflagration. Various times have been given for the completion of his move to Fort Mason, but it appears that he was well established there by sometime that evening. In his two accounts of the disaster, General Funston's actions after his move to the fort become vague. Aside from a few incidents touched upon here and there, some of which he had little involvement in, we are woefully ignorant of most of the orders he issued. If there are gaps in the record, it is because there are scarcely any records.

After Funston retreated to Fort Mason his behavior was not what might have been expected, given his actions in the past. Whether on scientific expeditions or fighting in Cuba or the Philippines, he had always opted to be in the midst of things, often risking life and limb in doing so. He was a courageous and heroic man, about that there is no doubt. But once arrived at the fort, he began to distance himself from the crisis, as if subconsciously aware that he might have mismanaged its earliest stages.

While Funston was at Fort Mason, men and women were being stopped on the street and ordered to pick up bricks or haul debris. While there is no civilian or military record of an order to impress citizens into service, many accounts relate incidents of their being forced to perform work. By observing the actions of the troops, though, we can deduce that Funston probably issued a general order calling for the impressing of civilians to perform work on behalf of the military—another exercise of authority by the army that far exceeded previous instances in American history.

Only General Funston could have decided to break with historical precedent, allowing a kind of temporary enslavement. People who were engaged in their own personal relief efforts or on a mission of some kind tried to refuse, but were threatened with immediate execution by bullet or bayonet. In the letter to her friend in the Philippines, Major Devol's daughter, Lucille, wrote of the impressed men, "Most were willing, but a few weren't. One, a chief of police from another city, objected to breaking stones and removing bricks, but he had to or be shot." The man Lucille Devol refers to was the police chief of Milpitas, a small town near San Jose. The number of men impressed during the disaster is impossible to ascertain, but it probably reached at least several thousand and could easily have been more.

CHAPTER 57

Sometime early on the first day, the California National Guard—the troops that would have been called had martial law been legally declared by the president—took up positions in San Francisco. Because of the absence of communications, they first operated without orders from their senior commanding officer, Adjutant-General Joseph B. Lauck, who was with the commander of the national guard in the Bay Area, Brigadier General John A. Koster, in the northern California town of Ukiah at the time of the earthquake, approximately 120 miles north of San Francisco. Command therefore devolved upon junior officers, usually at the company level. The guard units that mobilized were San Francisco's own First Infantry Regiment, the First Battalion of Coast Artillery, Second Company of the Signal Corps, and the cavalry's troop A. Over the succeeding days, they would be joined by national guard detachments from northern California as well as units of the cadet corps (today's ROTC) from the various nearby colleges and universities.

Koster and Lauck were finally able to reach San Francisco on the afternoon or early evening of the eighteenth, and Lauck set up temporary headquarters. (His official headquarters was in Sacramento, the state capital.) His first command post was established at the Occidental Hotel (home to Major Carroll Devol and a few other army officers), but the advancing fire forced him to move to the Pacific Union Club, which he had to abandon only a few hours later. He went to the Fairmont Hotel for several hours, until the fire once again became too dangerous, and he decamped to the North End police station on Washington Street. When that, too, became unsafe, Lauck established his fifth and final provisional headquarters across the bay in Oakland.

The actions of the national guard during the fire are extremely difficult to ascertain. General Lauck relocated so often during that first day that it must have interfered in his communications with Brigadier General Koster.

Koster, meanwhile, had sent the guardsmen to those areas where they were most needed. The army itself had taken the choice areas of the city to guard, such as Nob Hill, the downtown and commercial districts, and the pleasant residential districts to the west. With the navy's small contingent of men under the command of Lieutenant Frederick Freeman having assumed control of the waterfront, this left the working-class Potrero and Mission districts for the guard, though some units appear to have scattered throughout the city, filling in where there were few army patrols. (On the fourth day, General Funston would divide the patrolling responsibility in the city into thirds. The army had its assigned district, which included all regular military assignments; the national guard would remain in its territory, and the San Francisco police would be responsible for the last third. This geographic arrangement would remain in effect until June, when the army was finally relieved of its duty in San Francisco.)

Besides assisting in firefighting efforts in the Mission and Potrero districts, the national guard also helped in dynamiting the buildings in those areas of the city. Here, too, the dynamiting caused many fires because the buildings were almost all made of wood, and toward the end of that struggle, probably on the third day, Koster appears to have made the decision to allow his men to accept the help of local volunteers—reportedly, three thousand men. This quickly turned the situation around, and the conflagration in the Mission district was finally stopped on the twenty-first.

Koster met with Mayor Schmitz a few times, and it appears that some of the mayor's instructions to the national guard conflicted with orders given to the fire department and the army. Though Funston had welcomed the marines from Mare Island and other services into his jurisdiction of the crisis, he disavowed any responsibility for the national guard. That might have been the case because the actions committed by the guardsmen, especially on the first day when there were no senior officers on the scene, demonstrated a lack of discipline that surpassed whatever unprofessional behavior Funston could tolerate in his own troops—this at a time when professionalism in armies was never an issue.

The national guard of 1906 bore little resemblance to the trained and tested organization of today, which has served with distinction in Kuwait, Afghanistan, and Iraq. Among its ranks were many undisciplined young men. During the fire the national guard quickly gained a reputation for being the most trigger-happy of the various services patrolling the city. Sydney Tyler writes, "Orders had been given by General Funston and the Mayor to kill any looter on sight, and the vigilantes and a large number of the National Guards took advantage of this order to practice their marksmanship on men who might happen along and who in their excited condition might be said to be acting in a suspicious manner."

When Frederick Rowe discovered four men in the Crocker Building trying to break into the safe of his firm, the McCloud River Lumber Company, the guardsmen took custody of the four and then shot them almost immediately.

And mercy shootings, which began on the first day of the fire, apparently got out of hand with elements of the national guard, for some were performed without the victim's consent. The marine from the *Marblehead* reported that he saw the "soldiery shoot the people in the Windsor Hotel on Market Street. People were at the third-story windows trying to get out. The militiamen shot them in order to save them from burning to death. But these men and women didn't get a chance for their lives. The soldiers had plenty of blankets and could have tied them together and made a life net. They had a chance, anyhow, to jump, as it was only three-stories high." The marine continues his account: "This was only one sample of the lack of discipline and common sense among the guardsmen. I have been on service in several great disasters—at Baltimore, in the great fire, and at Galveston also—and I never witnessed such raw work on the part of the militia. They shot down men mercilessly and brutally—simply to make a record, I guess." The marine also recounted that a number of the guardsmen he observed were drunk.

In a letter, an army PFC named Kaff reports, "One man and one woman, I know positively were killed by the military. They were pinned down in the ruins caused by the earthquake and there was no possible chance to get them out. Fire was within 50 feet of them and the man asked a soldier to shoot him. They were both shot in order that they would not have to suffer any more than possible."

"Soldiers shot living beings to save them the torture of death in the flames," recounted Margaret Underhill, who went on to describe an incident near Sacred Heart College:

> Pinned beneath the structure was a man who pleaded piteously with the men who worked to release him. His head and shoulders projected from the wreckage. With his free arm he tried to help the workers by pulling at the timbers. His eyes bulged from their sockets. One by one they were driven back by the flames until only one was left, a soldier. From where we stood we could see the very timber that held the man down. . . . His hair and mustache were singed.
>
> "For God's sake, shoot me," he begged. His voice rose clear above the roar of the flames. The soldier turned and went back.
>
> "Shoot me before you go," the man yelled. The soldier turned quickly, his rifle at his shoulder. The rifle cracked, and the blood spurted from the head of the man.
>
> I covered my eyes and went on.

The military were hardly the only ones reported to be committing mercy killings. Edward E. Clark recounts how a similarly trapped man asked a policeman to shoot him. After much hesitation, the officer shoots the man twice, but in his nervousness, only wounds him. The man's brother then seizes the gun from the policeman and shoots the trapped individual in the head, killing him instantly. Reportedly, the distraught man also turns himself in to the authorities and is immediately released.

Funston writes that there "was no necessity for the regular troops to shoot anybody and there is no well-authenticated case of *a single person having been killed by regular troops*. If there is any lesson to be derived from the work of the regular troops in San Francisco, it is that nothing can take the place of training and discipline, and that self-control and patience are as important as courage." In writing this, General Funston apparently had forgotten that he had wired the War Department at 8:30 P.M., April 20th reporting that looters had been shot.

Reports of euthanasia were common, particularly the stories that proliferated around the tumultuous winds of the firestorm, where the time to save lives or to save one's own life was considerably reduced. The most comprehensive accounts of firestorms are found in the reports of the fire bombing of German cities in World War II, where bombs of magnesium, phosphorus, and napalm, among other accelerants, were dropped regularly on German cities to ruin them and to demoralize their people.

Dresden, then a city of refugees who were fleeing from the advancing Russians, is the most often cited example of the savagery of firestorm deaths, but in fact Dresden was the least bombed of the German cities. The allies dropped 67,000 tons of bombs on Berlin, 39,000 on Hamburg, 37,000 on Essen, 27,000 on Munich, and just 7,000 tons on Dresden. But somewhere between 40,000 and 50,000 people were killed in two days of the Dresden bombings by 800 British planes and 527 American heavy bombers. The fire destroyed about 75,000 flats and created a fire wind of more than 1,800 degrees Fahrenheit that was reported to have reached velocities of more than 100 miles an hour. The wind was so great that it was said to have created something of a venturi effect, sucking people into it like a whirlwind. The fire also consumed the oxygen, leaving people to simply collapse in the street.

In San Francisco numerous people had been trapped in the ruins when the fire reached them, but the fire never behaved in the way it did in the firestorms of Dresden. The oxygen supply was not depleted, and no winds were strong enough to lift people off their feet. The fires of the Bay City burned slowly, for the most part; only a few accounts describe a whirlwind, and photographs of the event testify to a slow but certain burning.

But still, no matter the speed of the burning, there was much wanton behavior.

Guardsmen and soldiers were even accused of shooting at men of the other services and at policemen. (One police officer reported that he was shot at four times, but fortunately was not hit.) A military cadet from Berkeley, a man by the name of Aten, was on patrol duty in the city when he was shot and wounded by a regular soldier while trying to clear out a saloon. On Thursday afternoon, at the corner of Eddy and Polk, this uniformed student was, along with three other cadet students, ordered to go into a saloon and grocery store and close it down. He reported that it was quite crowded and unruly, and continued:

> We had just got the crowd nicely moving when I heard a rifle shot outside . . . and I caught a glimpse of a soldier standing on the sidewalk on the Eddy Street side of the entrance. At the same instant another soldier on the Polk Street side fired and I was hit.
>
> Several more shots were fired, but went over the heads of the crowd. . . . There was no excuse whatever for this shooting. We cadets were clearing the store without having to resort to strong measures. No material resistance was being offered, and a random shot into such a crowd of men, women, and children could not be justified. It may be that the soldier who did the shooting was drunk. It may be that he was too excited, or too inexperienced. Anyway, I was shot, and people said that a civilian was also hit.

The bullet shattered three inches of Cadet Aten's thighbone, just below the hip socket. Another soldier picked Aten up and helped carry him to Jefferson Square, out of the crowd and approaching fire. No one thought to identify the soldier who shot, for the immediate requirement was to move the cadet, who was screaming in pain. He was taken to the Lane Hospital, but because it had completely run out of water he was moved to the Presidio hospital. Aten was brought into a surgical ward where there were fifty-four patients and just forty-four beds, including the one in which the fire chief, Dennis Sullivan, would die the next morning. Still writhing in pain, he was placed on a folding cot until a benevolent soldier got up from his own bed and insisted that the wounded cadet take it.

Aten continued in his narrative:

> Three patients suffering from gunshot wounds inflicted by soldiers were received while I was there. One of these was a chinaman who was a prisoner and tried to escape. He was shot through the head and subsequently died. Another was a Japanese that ran away when a soldier challenged him, and was shot through the hand, the head, and the shoulder. He recovered. The third was myself. There were seven hundred patients in the place, and as far as I could learn, we three were the only

> ones shot by the military. . . . The hospital was overstrained by the pressure of the first few days, the congestion of patients, and the lack of nurses, food, and water. I . . . did not get a bath until the following Monday. During that period my only nourishment was liquor and raw eggs. . . . By Monday, however, everything was ship-shape, and the hospital was running on lines of military precision.

One guardsman even tried to assassinate Jeremiah Dinan, San Francisco's chief of police, but was eventually judged to be insane and incarcerated in an asylum. After the fire, Mayor Schmitz pleaded with the army to leave its patrols in place beyond the date they were scheduled to return to their bases, but asked Governor Pardee to withdraw the national guard the following Tuesday because of numerous complaints.

These accounts are not intended to suggest that it was only the national guard that added to the disorder in the city. In the investigation of one looting charge that occurred near the custom house, the army questioned a Captain Orrin R. Wolfe, "Was there much drunkenness among your men during the period that you were on duty at the Custom House?"

"Very little indeed," answered the captain. "One or two men were under the influence; there was more or less drinking as was natural since the whole community was a barroom and liquor was flowing like water, but I saw no man in my company . . . that was so drunk that he could not walk, and only two or three that were not able to perform their duties at all times."

Even ad hoc groups took part in the vigilantism of these chaotic days. At the time of the 1906 earthquake, retired Colonel Thornwell Mullally was the executive assistant to Patrick Calhoun, who was the president of San Francisco's principal public transportation company, United Railroads Company. Calhoun was also the nemesis of Phelan and Spreckels, and a man who would become central in the corruption scandal.

After seeing the army troops enter the the city to supplement the police force, Mullally went to Mayor Schmitz to propose that United Railroads' two thousand employees, most of whom wore uniforms similar to the police, be used to help in the patrol duties. Schmitz readily agreed, giving permission to Mullally to muster his workers for this duty. According to Colonel Mullally's account, the United Railroad men were excited by the new responsibility they had been given. Although arms could not officially be issued to them, many quickly obtained weapons from their homes or from friends and were later accused of killing suspected law-breakers. It is not known if General Funston agreed to this use of man-power, but it is in any analysis another example of a misuse of the time and effort of men who could have been pulling hose or wet mopping the sides of buildings.

There is an ironic aspect to these accounts of the street actions during the disaster. Of all the men serving in a law enforcement capacity during the disaster, those least inclined to turn to the excessive use of force were members of the city's own official law enforcement agency: the San Francisco Police Department. Perhaps two deaths can be attributed to the police, but there is no doubt that the city's officers were not predisposed to killing individuals without trial or jury. There are many stories of policemen making difficult arrests during the fire, sometimes at great danger to their own lives, and they were generally reported to have performed their duties in a humane manner.

Police officers accustomed to walking a beat were well experienced in dealing with people who were weak as well as every type of malefactor prone to foolishness or error. Whenever they witnessed looting after the quake, therefore, the police were likely to have seen the looters as individuals responding to an overwhelming situation. There were instances of pharmacies' being broken into for medicines by medical authorities. In North Beach, sailors broke into a sporting goods store for guns and ammunition to use in controlling an unruly crowd. Laffler also reported seeing a group of fifteen or twenty firemen who, on the second day, had looted a store of cheese and olives. When he remonstrated with them, one replied, "Well, if you had been dragging the hose around for two days, and it hadn't done a damned bit of good, you'd be wanting some breakfast, too."

There was a general and desperate need for food by many who had become sudden and incognizant refugees, especially by the second day, when most relief supplies had only just begun to enter the city. Mothers needed milk for their children, and otherwise respectable men broke into stores to acquire it. Even the stealing of alcohol in at least one instance had a reasonable motivation when a man forced his way into a liquor store to alleviate the injuries of a family member. That man was, reportedly, gunned down by a soldier as if he had been one of the "lowest fiends" the soldiers had been instructed to shoot if caught breaking the law. San Francisco's policemen, however, long known for their tolerant nature, helped people as often as possible as their job required, and they saw the suffering population on their own terms, which is to say with understanding. So the police have survived history as true heroes.

CHAPTER 58

As Wednesday drew to a close the situation was growing progressively more critical. In small blazes, the general rule of thumb is that a fire will double in size every minute or so, but the growth of large fires depends to a great extent on the wind conditions and the amount of fire loading—how much furniture or stock is in a given building. Though there was little reported wind for most of the first day, a dangerous shift came in the evening hours.

The fires had not been spreading quickly but were nevertheless unstoppable—like waves coming in from the sea. Firemen in the southern part of the city were trying to hold against the Hayes Valley fire, but found it, block by block, an impossible task. At the same time in the northern part of the city the sheer heat of the blaze pushed the firemen back six blocks north of the start of the Chinese laundry fire. And a few streets west, and not more than a dozen blocks from there, Engine 1 and other fire companies went along Montgomery from hydrant to hydrant, and from cistern to cistern without finding water.

Three large fires north of Market had merged into one, and Jack Murray had to have been thinking about the fire's possibly encroaching Telegraph Hill and formulating some plan of action for his family to join the continuing stream of thousands of refugees seeking protection. He himself was now on Washington Street, where Engine 1 had dug in with just meager cistern water between them and the fire. Their immediate goal then was to keep the fire from taking the section of Montgomery Street known as the Monkey Block, one large building that was home to many writers and artists. This important historical structure began as an office building in 1855, funded with an investment of one million dollars, a huge sum of money at that time. A four-story building designed in the Italianate style, it was an engineering wonder of the mid-nineteenth century: To keep it from sinking into the made land of its site it was constructed to float on a redwood raft that had been sunk into the ground. This not only gave it stability but also allowed it to float during the tremors of earthquakes. The Monkey Block was a haunt of many notable writers of the time—Robert Louis Stevenson, Bret Harte, Ambrose Bierce, and Mark Twain. It was later converted to residences and was at times home to Diego Rivera and Frida Kahlo, Jack London, George Sterling, Frank Norris, and Margaret Anderson. As New Yorkers have mourned the loss of Penn Station, San Franciscans will never forgive the shortsighted decision of the city's fathers to allow the building to be razed in 1968 to create space for the pyramid of the Transamerica Building.

Just across the street from the Monkey Block was the U.S. Appraiser's Building, a four-story Victorian redbrick edifice built in the industrial style. Known as the customs house, the massive structure had been designed in 1874 by the federal architect Alfred Mullet. Murray took notice that it was only a large excavation that had been made to build the foundation for a brand-new Eames and Young customs house that was keeping the flames from the Appraiser's Building. Because of the dislocated pipes in the area, that excavation was filling with the precious water that would later be used for drafting by the firemen.

A short distance up the street from the customs house to the west, on Kearney, stood the hall of justice, just ten years old and built on the site of what were once notorious gambling dens. It had also been the site of the old Jenny Lind Theater, which some years earlier had burned to the ground. And now the huge conflagration of San Francisco, as it turned up Clay from Montgomery, was determined to deliver fire once again to that very location.

It was little known that the Monkey Block also housed the rare books of the magnificent Sutro Library. Adolph Sutro was an immigrant Jew from Aachen, an entrepreneur who had made a mining fortune from the Comstock Lode. He was also an engineer who had been popular enough to become mayor of the city in 1894, only to be defeated by an even more popular James D. Phelan. But Sutro believed that his most significant accomplishment was the library that had been inspired by his profound interest in books. In two buildings he kept a collection of more than 250,000 to 300,000 volumes, which was considerably more than the holdings of the entire city library system. Indeed, it was believed to be the largest library in single hands in the world—no small testament to the intelligence, perhaps the genius, of Sutro, who was surely the most important bibliophile of the second half of the nineteenth century. While many of his books were held in a building at Pine and Battery, the most notable items in his collection were all stored at this Montgomery Street location.

The prizes in the Sutro Library were many, including four complete original Shakespeare folios, one of which, ironically, even had singed edges from the Great Fire of London of 1666. There were also Shakespeare's rent rolls for Shottery Meadow in Stratford, records of his real estate holdings, and copies of his sonnets.

There was also an authenticated Psalter that had been placed in the hands of Charles II as he was restored to the throne of England in 1660. This Psalter, *The Psalms of David*, was printed in London by Robert Barker in 1615 and was originally owned by James I, the first Stuart king. There were several originals of the Book of Common Prayer, original folios of Ben Jonson, original Gutenberg and Caxton printings, old Hebrew scrolls dating back to the tenth century, and, most important to Californians, a very large collection Sutro had bought from the Mexican government of "... books, diaries, and manuscripts bearing upon the early history of California and lower California, especially the mission period in the boundaries of the present United States."

Sutro also owned America's single most important scientific book of the eighteenth century—Benjamin Franklin's *Experiments and Observations on Electricity Made at Philadelphia in America*, published in London by E. Cave in 1751, which included a supplement that described the famous kite and key experiments. Sutro had also collected more than three thousand incunabula, books printed before the year 1500.

It is probable that Jack Murray and the Engine 1 firemen knew of the Sutro Library, for the Monkey Block building was in their first-due district, which meant they would be the first due to arrive at any alarm of fire there. All firemen take pride in knowing the special conditions of buildings in their first-due district. From a purely technical standpoint, any building that held that many books would certainly draw the attention of a fireman, not only because of the flammable nature of paper but also because the heavy floor loading of books influenced the possibility of collapse.

Chief John McClusky of the 1st District was on medical leave but rushed into the city from across the bay as soon as the tremors stopped. He fought the great fire for the most part in the downtown and Chinatown sections of the city, and he reported that he and his men had worked for hours at the Montgomery block and at the federal Appraiser's Building, using whatever water they could find, whether from cisterns that had not been filled up with debris or from excavation pits. "We used water sparingly," he wrote, "distributing it as much as possible to prevent the buildings from catching fire." He and the men of Engine companies 1, 20, and 30, with the assistance of a few firemen from Oakland, fought as hard as they could until they were joined by Lieutenant Freeman, Midshipman Pond, some of the crew of the *Perry,* and a few straggling soldiers. This larger group held its position here and kept the Montgomey block as wet as it could until the south side of Washington Street had burned through. Jack Murray, Captain Murphy, and the rest of Engine 1 were by now physically and emotionally spent, but they continued searching for water and feeding the hose whenever they found it.

Then a book lover's miracle: Whatever breezes were in the air had suddenly shifted. The fire, to the amazement of Jack Murray and the firemen of Engine 1, did not continue north up Montgomery, but turned to the west, missing the one hundred thousand wonders of the Sutro Library by a matter of yards. All books are valuable for one reason or another, and when they are original pressings or hand inscribed they can become precious. Though Sutro's most valuable books were saved, much of the Sutro collection—more than two hundred thousand volumes—was already burning as the fire made its way behind the firemen along Battery and it reached the second warehouse of the Sutro Library. All that remained of his collection, including fewer than fifty volumes of incunabula, can be found today at the Sutro Library branch of the California State Library.

Chief McClusky then realized there were no firefighters available to fight the hall of justice fire, and he pulled his men from Washington Street. Jack Murray and the firemen did their best when they arrived there, but the fire was too great and the water lines too thin, so in great despair they watched the building burn to ashes. None of them realized what a victory had been achieved when Providence had preserved the books that

Adolph Sutro had had the intelligence to gather as vessels to a civilized future.

CHAPTER 59

Perhaps no account of a single fire in the great conflagration tells as complete a narrative as the saving of the U.S. mint on Mission and Fifth, just one block east of Market. Built in 1874, it was the second largest mint in the nation, and since paper money was rarely used west of the Mississippi, it was the depository for about two hundred million dollars in coins and gold, an enormous sum (about four billion dollars in today's value). Its superintendent, Frank A. Leach, a former newspaperman, lived in Oakland, and in his subsequent report to his Washington superiors he related that he first thought that the earthquake was a problem local to Oakland, one that threw off the chimneys and chipped the masonry of every house he could see. He was startled when he finally had a clear view of the city across the bay, as he saw "that the heavens above the city were filling with the black smoke of a great fire," and realized for the first time that San Francisco was burning.

Though all traffic to the city had been prohibited, Leach immediately made his way across the bay, having first gotten permission from the head of the ferry division to leave the port. He had to triangulate his way to the mint because the army had closed so many streets, and he commented, "Upon reaching [Powell] I saw soldiers along the thoroughfare to keep all people from passing into the burning district. Just what advantage to the public, property owners, or anyone, for that matter, such use of the soldiers was, or of what value their instructions were, I could never learn or understand."

The soldiers would not let Leach pass over Market Street, either, despite his protestations and identification, and it is only through the intercession of a policeman who recognized him that he was able to cross the street.

About fifty employees as well as a small detachment of soldiers under the command of a Lieutenant Armstrong had reached the mint before the soldiers barricaded the streets. Earlier, General Funston had sent a detachment to every federally owned building in the city, including the post offices and customs house, and ordered them to limit their involvement to guard duty.

The mint building was a squat, stone edifice in simple Doric style, three stories high, with a plain pediment above the main entrance. It was more than two hundred feet wide, and two thirty-foot chimneys towered above its roof. Superintendent Leach immediately organized his employees into a firefighting force, distributing them throughout the building, at the windows and on the roof. No city firefighters were present, for the men of Engine 17, whose firehouse was just across the rear alley at 34 Mint Street, had responded much earlier to the electric works building fire at 3rd and Jesse, just two blocks away. In any event, the superintendent did not want any but authorized personnel on the premises because of the exposed treasure. An artesian well had been recently installed for firefighting purposes in the building's yard, and it proved to be a fortuitous addition for it allowed the pumping of two hose lines. Leach was beside himself, however, when he discovered that the shaking had completely ruptured the pump.

The fire was now approaching slowly both from the east, where all the buildings were mostly small and had much less furniture, files, stock, and other fire loading within, and from the west—that is, from the south side of Market Street—where the buildings were much larger. In a letter to his brother, a mint employee named Joe Hammill told of flames that were "200 or 300 feet in the air." It was undoubtedly as critical a fire environment as existed anywhere that first day. The employees watched patiently as Leach supervised the reconstruction of the pumps. Every second counted as the fire drew closer and closer, taking first the Metropolitan Temple, then the Lincoln School, and then the great Emporium Building, just across from the mint. Engine 17's firehouse, built in 1888, was now burning as well, adding to the wall of flame attacking the mint building.

Finally, the pump was realigned and began to churn as the fire came in from the west and the northwest. And it came furiously, first in heaps of smoke that made every room dark as night, and then in great waves of flame primed by the large buildings on the west side. The hoses were brought to the roof, and while the building would have been defenseless without them, they were not strong enough on their own to hold the fire. As Leach wrote, ". . . these alone would not be sufficient to keep the fire from gaining a foothold. On the second and third floors the men worked almost wholly with buckets. Every man stuck to the post where he had been placed. There was not a whimper, though . . . all felt . . . the issue of the contest with the great mass of fire that was soon to sweep against us."

These were courageous men who felt themselves in the presence of inspiring leadership. As Joe Hammill wrote, "[Leach] took his turn at the hose, and did not ask his men to go where he would not go himself." Showers of flaming cinders kept falling on the roof, igniting it again and again, and the men would extinguish the flames and pull the roof boards up, wetting down the beams as they became exposed. It eventually got so hot there that the roof team was forced down the stairs and was trapped between the fire above them and the fire in the street. With headbands around their mouths and coughing in the smoke, they made a valiant push forward, attacking the fire at every window and every doorway until they again found themselves at the roof, yelling in victory. The machine room on the third floor was entirely filled with fire, but they threw bucket after bucket at the flames until they began to die down. The roar of the firestorm about them and the continual sound of explosives and falling buildings were terrifying, but they fought on. It took them more than five hours of relentless, determined struggle to keep the building whole, and when it was over, at about 5 P.M., Leach gave them leave to look after their own homes and families.

Just two blocks south was the new U.S. post office building and courthouse, completed only the previous year after twelve years of construction. It was a magnificent building, perhaps the best example of the Beaux Arts classical style in the United States, the crown jewel of the city's architecture. As the fire came at it in waves from the west, the post office staff tore down every curtain and pushed every stick of furniture to the center of its many rooms and then stationed themselves at every window. The guard soldiers posted outside never entered, and so it was left to the ten employees of the post office to risk their lives to save the building. They fought for two hours, wetting the walls and windows by smacking at them with saturated mailbags and throwing pots of water drawn from the hydraulic elevator system. The structure had suffered significant earthquake damage, and its foundation had to be rebuilt, but the gallant postal workers lost just two rooms to the fire, and both were perfectly restorable.

It was another island of preservation in the midst of destruction, and it is no accident that so many key federal buildings in the city did not burn while other stone and brick buildings did. The army had no authority in these buildings and so could not order them evacuated. The Ferry Building, the U.S. post office and courthouse, the U.S. mint, and the U.S. customs house represented telling examples of how men, when left to their own devices, could defeat even a fire as great as that which confronted them on that day.

CHAPTER 60

Annie Murray, wearing an apron that fell to the bottom of her ankle-length wool skirt, walked her children to the Filbert Street side of Telegraph Hill to view the fire. The youngsters were awed by what they saw, a view that would be etched in their minds forever. "Can I go down and help the firemen, Ma?" young Will asked. The Murrays were a firefighting family—Annie's brother was a fireman in Oakland as well—and Will believed he had a firefighter's right to be near the action. His father had so often let him sit around the firehouse and occasionally let him polish the tools and the big brass bell that floated just beneath the bucket seats of the steamer. But it was a dangerous day. "No, son," Annie said with the strength of voice that she knew would keep him from asking again. She brought him close to her apron while his three sisters circled their mother. They looked out over the steeple of St. Francis Church just off Montgomery, where Annie and Jack had married in 1890. Josie, womanly even at fourteen, held the baby, Essy, in her arms. Annie sighed as she watched the smoke being carried westward by the wind. The fire was expanding. Jack had promised to take her to a musical at the orpheum that Saturday on his one day off in a month, and now he might be gone all week, with the size of the fire before them. The children pointed to different landmarks. The Palace Hotel was in the path of the fire. They could see that a section of the city hall dome was still perched on a structure that had mostly fallen away. But at least the fire was a safe distance away, and she comforted them all. It would not, she assured them, come up Telegraph Hill. But as Annie watched the city kilning and crumbling before them, she added, "Just look. You children will be the new pioneers of San Francisco."

The following morning the Murray family again left their home to gaze out over the burning city. Looking toward Davis and Jackson, they saw a peculiar, if not wonderful, sight. Several freight cars had pulled onto a siding next to the Swift Packing Company and inside were thousands of chickens. They watched as a couple of sailors opened the freight cars, and the street filled with flying, fluttering poultry. Some boys ran to the street and began to chase the birds, as much for sport as for dinner. Will again begged his mother for permission to run down the hill to join in the excitement, but Annie held fast to his hand.

CHAPTER 61

The blasters were setting dynamite on Montgomery Street, and Edward Lind, the manager of the Hotaling Whiskey Company, which was between Jackson and Washington and one block from the Sutro Library, talked Captain Orrin Wolfe of the 22nd Infantry out of blasting his warehouse. He had advised the officer that it held so much whiskey that the resultant explosion would take down the U.S. Appraiser's Building across the street from his rear entrance. The soldiers considered this warning, and then left the whiskey warehouse to turn their attention to other buildings. The warehouse never did burn and so gave rise to a popular doggerel that asked why, since San Francisco was known as such a wild town, God allowed the churches to be destroyed while saving the Hotaling Whiskey warehouse.

In the meantime, Lind brought together and hired a group of eighty men to roll the whiskey out of the building. They deposited twelve hundred barrels on Battery Street and waited there to be paid. Lind had no money because his company safe had been unsettled and jammed in the earthquake, so he look a ferry to Oakland to borrow from his parents. When he discovered they were not at home, he took a loan from a local barkeep. Looking for a ferry to return, however, he found that an order had come from San Francisco prohibiting anyone from crossing the bay to the city. Lind then had to search for Governor Pardee, who had established an Oakland office, to plead for a return pass. Finally, the governor relented, and Lind was not only able to pay his moving men, but also to secure guards for the next three days until he could redeposit the barrels in his warehouse.

Lind's cache of whiskey was probably the only whiskey in San Francisco that survived. Upon a general order from Funston and also from Schmitz, all the whiskey in the city had been destroyed, a measure that had been intended to prevent drunken disruption. Because Captain Wolfe had been focused on dynamiting the Hotaling warehouse, he paid little attention to the liquor itself, so all the Hotaling stock was preserved.

CHAPTER 62

At about 5:00 P.M., the fire consumed the hall of justice. The bodies taken from the morgue were still being buried in Portsmouth Square, and the city's vital police and courts records were still being covered with wet blankets when the fire began to push everyone back. The part of San Francisco that could have been called the village in the early days of the city was now a smoldering ruin. The dynamiting crew was just ahead of the fire as it traveled up Clay, but they realized they were out of dynamite when they decided to blow a building at the corner of Kearney and Clay, the borderline of Chinatown. Instead, they used black powder, a powerful and highly flammable explosive, and as the building exploded, it sent burning mattresses, chairs, sofas, and kindling flying across Kearney Street, setting the fires that would doom Chinatown. Before midnight, more than fifteen thousand Chinese would be forced from their homes. Men carried boxes in their arms and on their heads. Women whose feet had been bound and who wore embroidered three-inch platform shoes found themselves in public for the first time since their youth, teetering along in a frantic movement of people.

Donaldina Cameron rushed her little charges through the crowded streets, keeping careful watch. There was no telling if someone, even in this horrendous crisis, would try to snatch one of them. It had happened so many times before: Someone would try to "reclaim" what he believed was his property, a young girl bought and paid for. These children, most of them, had already been kidnapped and enslaved once, and with God's help she would get them through this crisis safely. They went to the top of the hill from 920 Sacramento and then down the hill to Van Ness. Even now she was thinking she had to change the name of her mission house to Mei Lu Yuen, the Garden of Beautiful Family Relationships. Her children, each of them, deserved a family and love as much as food and shelter. She was taking them now to the only place in San Francisco where all fifty of them, each carrying her essentials, might be taken in—the First Presbyterian Church, sometimes called the Old First Church.

Once they arrived the minister and his wife welcomed the small and frightened army and distributed them among the pews so that they might rest their heads against their bundles. Donaldina, seeing her charges so well cared for, thought to return to her mission to retrieve the papers that conveyed the history of each child, documents that might be important to their futures and which she had neglected in the rush to vacate. She wrote,

> Martial law had cleared the desolate streets of all living things for many blocks. Thanks to one soldier's sympathetic heart, we passed the closely guarded lines, and were permitted, with many warnings to make haste, to enter our Home. The red glare from without lit up each familiar object in every room. . . . On the block below a terrific blast of dynamite was set off. The soldier on duty outside imperatively ordered us to make haste. We gathered papers and valuables . . . took a final look through the dim shadows . . . then a last good-night to [our home] . . . two hours later, the flames had wrapped it round.

CHAPTER 63

The fire continued to rage on Market Street, and as the Spreckels Building stood burning, soldiers carried boxes of dynamite into the new Monadnock Building next door. Because it was of brick and steel construction, however, no matter how many times they blasted it, the walls simply blew out while the building itself remained strong. Soon, just north of it, the Palace Hotel caught fire.

The Palace was the city gem most beloved by San Franciscans. When Selim Woolworth sold the property for the hotel to the banker William Ralston in 1868 he is said to have exclaimed, "I don't like earthquakes. I'm getting out of here." Ralston then invested five million dollars and went on to build the largest hotel in America—a seven-story structure with eight hundred rooms and the most up-to-date fire protection appliances and system in the world. It was much bigger than the city needed, but Ralston was optimistic about San Francisco's rapid growth. Ralston had founded the Bank of California and invested heavily in the Comstock Lode and in the building of Virginia City, where Mark Twain would cut his journalist's teeth. He himself had come from Ohio and like Twain had worked on the riverboats of the Mississippi. Ralston was a true and honest entrepreneur. At six feet tall, with one of the handsomest faces in the country, he was unfailingly admired, and as one of the wealthiest of San Franciscans he was naturally one of its paramount social leaders.

He had also invested in the Palace to create a magnet building to develop a large expanse of property he held South of Market. He also owned the Mission Woolen Mills, the Cornell watch factory, the Kimball Carriage Company, the West Coast Furniture Company, a sugar refinery at a time when Claus Spreckels was still in groceries, the Hunter's Point Dry Dock, and the Grand Hotel. He owned an expanse of property on Montgomery Street that he developed in the 1860s. He tried to develop property south of the city thirty years before Burlingame became the country-home town of choice for well-to-do San Franciscans. His plans were ambitious and many, and he determined to make his holdings and his city among the most important in the world.

He spared no expense in making the Palace fireproof. Iron banding was placed around the brick walls to increase their stability in an earthquake, and the roof was covered with iron to protect it from fire. A dozen water hydrants were situated around the building at curbside to assist the fire department, and a cistern of 358,000 gallons of water was built in the cellar as a reserve. Four artesian wells were dug beneath the hotel, and a piping and pumping system could relay water anywhere in the building—with carts of hose placed on each floor to be used in firefighting.

When the Palace Hotel took fire, however, the employees became overwhelmed with uncertainty. Though they had access to cisterns, water pumps, and hose, they had no training in the fundamentals of firefighting. Their instinct directed them to find a hose with water and put the fire out, but not realizing how precious their water supply was and how delimited, they decided to put the strength of their system to wet down the roof to protect it from the embers that were flying like falling snow from the eighteen stories of the Spreckels Building. By the time the firemen arrived, very late in their response due to the plethora of emergencies all around them, almost all of the hotel's water supply was gone. Even the hose that one employee had attached to a fire hydrant immediately fronting the entrance to the hotel was removed by the city firemen, who connected it to one of their steam pumpers—a connection that served the firemen fighting the many fires several streets away, down toward the Ferry Building. By dinnertime on Wednesday the largest and perhaps most elegant hotel in the country was doomed to destruction, because every fire contingency had been considered except the possibility of an entire city's burning around it.

CHAPTER 64

The Committee of Fifty finally came together at the Fairmont Hotel at about 8:00 P.M. It was expected that the main order of business was to announce the important committee memberships and work assignments. But before this took place, the assembled men and women were given a presentation by General Funston. It is obvious that he had been traveling through the city, as he was described by attendees as being "grimy and red-eyed from the smoke."

Funston stood before them, very much a man who was accustomed to command and authority and advised the group as to his idea for the tactics of fighting the fire. "Dynamite is the only answer," he insisted. There was no fire department official at the meeting—in fact, not one person in the gathering had any firefighting experience—and so the general explained his plan with a forefinger that traced a line across a map of the city.

Though the army had been woefully short of high explosives that day, Funston reported that more could be obtained from powder companies and army garrisons throughout California. "In fact," he said, looking at his watch, "the first supplies should be here within the hour."

His proposal met with no objections and no alternative solutions were suggested. Funston had already met with Schmitz and Phelan, Chief Dougherty, the dynamiters Bermingham and Briggs, and no doubt other members of the committee earlier that day. It seemed evident that he had the support of everyone.

"There is no water," Funston continued. "The only hope I have of slowing up the advancing fire is by using dynamite. A fire wall can be blasted that will hold the flames until water can be provided." Not one person in this well-educated crowd remembered the disappointing experience with dynamite during the great Baltimore conflagration of just two years earlier, despite the fact that it had been widely reported in all the nation's newspapers.

When additional water might be expected or from where it would come was not explained by Funston. It was obvious to all present that with the water mains broken, it would not be arriving soon.

Uneducated in fire control tactics, Funston did not propose using the last broad avenue, Van Ness, as a firebreak, but instead suggested a position approximately six blocks to the *west* of it, as the hydrants seemed to be working on the other side of Webster Street. Fortunately, as the days progressed, the fire department was less willing to consign so much of the city to the flames, and a stand would be made finally at Van Ness, though it would certainly prove to be a touch-and-go fight of a lifetime.

After a long pause, Schmitz conceded publicly to the dynamiting plan, and said, "Very well. We are to be kept informed of each building to be destroyed." His order was issued almost halfheartedly, without any expectation of compliance. Essentially, the committee had agreed to what was already a reality: The U.S. Army would be making the critical decisions, though supposedly in consultation with the civic authorities.

On a more positive note General Funston reported that the army had already taken the lead and had ordered tons of food and thousands of tents from the war department. Actually, every spare army tent in the whole of the United States was being shipped to San Francisco. Those tents would barely be sufficient, for even considering the many who had fled the city, two hundred thousand refugees needed shelter. The crowd applauded as General Funston took his leave, and indeed, the army did deserve much praise for the concerned and efficient way they had provided potable water, clean facilities in the refugee centers, orderly inspections to maintain standards of health and community in the tent camps, and also for the intelligent and humane effort that went into staffing and equipping temporary hospitals for the suffering, providing decent nourishment for all the displaced families, and protecting them from scoundrels and malefactors.

As the crowd dispersed Rudolph Spreckels saw his father and greeted him with a heartfelt handshake, as much a manifestation of affection between a father and son in public as the times would permit. In attendance, Rudolph noticed, was A.B.C. Dohrmann, of the then famous Dohrmann's Department Store; M. H. de Young of the *San Francisco Chronicle;* silver king James Flood; and bankers Henry Crocker and O. K. Cushing. Frank J. Heney, the respected prosecutor, recently returned from his work in Oregon and Washington, had also attended. There were two Sutros on the committee, Gustave and Charles, as well as Frank and John Drum, J. Downey Harvey, Tirey L. Ford, Thornwell Mullally, John Martin, and Clem and Joseph Tobin—among the most powerful of San Franciscans in wealth and social status. Just about every consequential citizen of the city had been asked to serve, though it is worth noting that Drum, Ford, and Martin were about to be indicted in the corruption prosecutions. The accusations made against them—peers and friends of many in the room—would bring about a reconsideration among the city's business class of the wisdom in continuing any corruption investigations, which were perhaps threatening to fall too close to home for the peace of mind of many of them.

Schmitz was the chair of the committee, and with the help of the lawyer Garret McEnerney and his millionaire friend J. Downey Harvey the subcommittee chairs were determined and assigned. Phelan was named head of the all-important finance committee, and he made certain that Rudolph was assigned it as well. They would have authority over the allocation of all relief funds, including the ten million dollars that would come from the U.S. Congress. This money would be delivered through the trusted offices of the American Red Cross and would be used for immediate relief efforts as well as for the rebuilding of the city.

It was obvious that the mayor had chosen to lay aside any personal grievances because he had chosen three of his leading political foes in Rudolph, Phelan, and Heney. Phelan made a reference to "earthquake love" to his friends, and Rudolph and Heney agreed with that sentiment. It was a time of crisis, and all San Franciscans would have to come together to relieve the pain of their fellows and families.

Besides Abraham Ruef and every member of the board of supervisors, Patrick Calhoun, president of United Railroads Company, was also passed over, probably because he was not a San Franciscan and perhaps because he was so closely connected to Ruef.

It had been whispered among these men that significant investigations had been undertaken by Prosecutor Heney, and it was as if the mayor had suddenly decided to cut the puppet strings of politics that tied him to the Labor Union Party, and to strike off on his own. The newspapers celebrated the courage of his singular command, and even Older's *Bulletin* ceased its daily attacks on the mayor's integrity. In the days after the earthquake, the business, religious, and civic leaders would give Schmitz at least temporarily high marks for his independence and emergency leadership. And Schmitz, in later testimony, would say, "My life started with the earthquake." Even his harshest critics acknowledge his service in the first days of the earthquake.

In a catastrophe a leader, no matter how tarnished his reputation, had to be supported. No one commented upon the influence General Funston had upon a mayor of essentially weak character and on his decisions, though everyone praised the general for his role in keeping order in the city.

Schmitz's independence did not last long, however, for Ruef later complained about having been left out, and he was given the chair of a committee for the permanent relocation of Chinatown, an idea that caused so much controversy that it never advanced. The real estate interests of the city saw an opportunity to make better use of the now-emptied streets of the area, and it was first suggested that the Chinese be moved to the fringes of Fort Mason, and then south to Hunter's Point on the bay, bordering San Mateo. The suggestions smacked of racism, and the empress dowager of China made her views known to President Roosevelt, a communication that effectively put an end to any talk of confiscating the Chinese property in San Francisco.

Then in the first week of May, Ruef again became entrenched in the rebuilding process when Schmitz created the Committee of Forty, from which he removed Heney and placed Abe Ruef as the chairman of the most important committee—the committee on committee membership. An outraged Rudolph, seeing this appointment made to a political boss who was determined to control and profit from the rebuilding process, resigned from the committee immediately.

In fact, Prosecutor Francis J. Heney, who had informally already begun to investigate corruption allegations, had reported to Phelan and Rudolph just that week that he had found hard evidence that the mayor had taken significant bribes in licensing "French" restaurants, and the investigations of other instances of corruption, known or alleged, were going forward with dogged determination and control. Previously, without any success, Rudolph and Phelan had attempted to get the support of other wealthy leaders of the city in underwriting the investigation, but not one was willing to publicly raise his head in moral indignation. It was safer when doing business in San Francisco to remain silent, and so Phelan and Rudolph were left stranded on an island of conscience, isolated from those of comparable education, money, and cultivation.

CHAPTER 65

As soon as the fire was controlled, Phelan decided to incorporate the finance committee so that its activity would be transparent and legally accountable. The Red Cross section had meanwhile consolidated with the Citizens' Committee, so there was every reason to believe that the relief contributions would be prudently handled. William Franklin Herrin, a public relations expert for the Southern Pacific Railroad, however, who was on Phelan's committee, was against incorporation and suggested it would not be warranted under the law.

But Phelan prevailed, and Rudolph Spreckels, A.B.C. Dohrmann, I. W. Hellman, Charles Sutro, and sixteen others were appointed to be responsible for the money, with Phelan as president, Dohrmann vice president, and Herrin second vice president. Its leadership was a politically balanced triumvirate: reform politics with Phelan, trustworthy business with Dohrmann, and business as usual with Herrin.

The committee called a meeting for advisement and invited the city's religious, political, and financial leaders, including E. H. Harriman, to attend. The amount of money that was expected—more than ten million dollars (two hundred million dollars today)—was announced. Harriman, who did not know Phelan other than as a former local mayor, suggested that a treasurer be immediately elected, preferably a banker, and that the contributions be spread among many banks throughout the country. Herrin agreed with his employer and moved that Hellman be nominated as the treasurer.

About this, Phelan later wrote:

> . . . It was obviously a prearranged plan to get possession of the funds. I ruled the motion out of order on the ground that

this was a conference, and that whatever action desired to be taken should be taken by the Finance Committee; whereupon, Mr. Harriman and his following left the room, and the meeting broke up.

President Ide Wheeler of the University of California said that I had offended Mr. Harriman; that he was very angry, and that it was a mistake to have alienated so good a friend to the community. I told him that Mr. Harriman had no reason to be angry, and I would speak to him. I went . . . [from the Hamilton School] to the lawn, and found Mr. Harriman scowling and black with a hostile expression on his face. I explained that, in the first place, the meeting was one for conference, to suggest ideas, but not to take action, as action must be taken by the Finance Committee; and, in the second place, that as chairman of the Finance Committee I was custodian of the funds, which were deposited in the United States Mint, and that permission had been so granted by the Treasury Department, and I could not consent to the transfer of the funds to a treasurer without approval of President Roosevelt, who, in his proclamation directed that all funds be sent to me, and that I felt a sense of great responsibility.

I further stated that I agreed with his idea of depositing the money in many places; that I was a banker, the President of the Mutual Savings Bank—and could do that very properly.

Mr. Harriman was greatly astonished when I told him that I was named in the President's Proclamation, and he forthwith relented, and blamed those who were about him for having failed to inform him of these facts.

On this occasion he said, "I am a friend of President Roosevelt, and have constant communication with him by telephone between our offices."

The incident passed, and there was no change in the organization['s officers]. Somebody observed that, "This was the first time anyone ever crossed E. H. Harriman and got away with it."

Phelan would face a similar challenge six weeks after the fire. It was well known that he had scheduled a trip to Washington to ask the

Congress for additional help in rebuilding the city. On the day before Phelan's departure, the mayor had scheduled a lunch with him and the members of the finance committee, during which he conveyed to them that the board of supervisors was insisting, since the immediate crisis was past, that the committee was now usurping its authority. Furthermore, since the relief money belonged to the people of San Francisco, it was right to transfer it so that it fell under the purview of the board of supervisors, the elected representatives of the citizens.

Phelan realized that his committee was being set up for another assault while he was out of town, so he immediately canceled his travel plan and continued his personal overseeing of the relief funds. The objections of the board of supervisors were not heard again because Phelan was too powerful a chairman to attack.

CHAPTER 66

Early Thursday morning at about 3:00 A.M., Lieutenant Freeman, accompanied by his junior officers Bertholf and Pond, boarded the *Leslie* and made an inspection trip from wharf to wharf. The three had had no rest since they stepped foot on the San Francisco peninsula the day before, and had already, with their men, fought fires at Steuart Street; at Meigg's Wharf, where the flames had come perilously close to the hundreds of refugees gathered at the water's edge that afternoon; at the Montgomery Street conflagration, where they had assisted Engine 1 and the other fire companies; and finally at the railroad sheds on Townsend, where they had single-handedly stopped the fire from destroying the railyards. Now, for the first time, they saw that the fires along the waterfront had been reduced to occasional puffs of smoke and small flare-ups that the fire department had under control, and they sought to rest. But as they looked into the city they could see about a mile away that another body of fire had burned through Chinatown and was now following a likely path up Nob Hill. Freeman decided to forgo sleep and find out if he and his crew could be of any use there.

As Pond took charge of the tugboat, Freeman and Bertholf half ran past the fleeing people to the perimeter of the flames, where they discovered that the firemen had been forced to abandon their one hose line when the fire overran them on Powell, between Sutter and Pine. Much of the firefighting force was now concentrated along Market Street: in the Mission district, where the Hayes Valley fire had traveled, and at a stand at the western perimeter of Chinatown. But the small amounts of water they had found in the cisterns, the excavations, and even the sewers had been quickly expended, and there were now no working hydrants along the Powell, Washington, Clay, Sacramento, California, or Pine Street hills. The firemen of San Francisco had no choice but to follow the fire as it traveled west and south.

On Powell, Freeman found a marine captain named Marix leading a detachment of naval sentries from Goat Island who seemed to be attempting to protect homes and stores from the approaching flames. In the meantime, more fully armed soldiers and marines had arrived, under orders to supplement the military personnel who were already evacuating and patrolling the city block by block. The roar of the dynamiting was continual, and Freeman reported seeing men blown into the air, such was the haphazardness of the operations. Later, it would be found that the drunken Bermingham, unfortunately the only well-trained dynamiter in the city at the time, was responsible for much of this wanton use of the ordnance.

Unable to be of any use at Powell, Freeman and Bertholf returned to Midshipman Pond and the *Perry* crew, who had taken a respite in the absence of their senior officers. Freeman ordered both the *Active* and the *Leslie* to Goat Island for breakfast and to take on additional water. But by then the water on Goat Island was nearly depleted, and they were allowed enough only for drinking with their meal. On leaving the island Freeman took twenty rifles and ammunition belts, which would be used over the next two days to patrol the waterfront area, where thousands of refugees had gathered to take advantage of the free ferry service offered by Harriman to escape the smoking city. No soldiers had been assigned to the wharves, the roughest section in town, and the seamen would attempt to make them safe for the city's fleeing hordes.

During this time the navy made an additional, little remembered contribution to San Francisco. In times of need, certain men move mountains to be of assistance, and such was the case with the chief electrician of the submarine USS *Pike,* James Curtin. Additional medical care was greatly needed, Curtin saw, and he obtained permission to commandeer what was known as the Old German Hospital, on Noe Street near Duboce Avenue. There Curtin established a medical center, and after a few hours of relentless supplication and plundering of abandoned stores, he was able to equip it with beds, linens, food, and medical supplies. He also recruited enough nurses and doctors to efficiently handle the many cases of emergency medicine that, as soon as word spread of an alternative hospital, began arriving at their door. The spontaneous infirmary became known as the Curtin hospital and not only continued its work through the terrible days of the fire but remained open afterward. Today it is part of the Pacific Medical Center.

Freeman noticed that the Oakland ferries had started bringing across the bay hordes of sightseers who had paid high fees to tour guides to see the fires burning at close range. They were a great interference, and he quickly ordered that all ferry traffic from Oakland be on an emergency basis only. Since many reporters had been scheduled to take these trips, Freeman and the military were later accused of having tried to prevent the exercise of a free press during the crisis. Eventually, passes were required to enter the city, which Freeman was authorized to sign, but, as Midshipman Pond wrote, "Freeman was a hard man to find. . . . He was not the type of man [who] would wait for instructions before taking action in an emergency."

CHAPTER 67

On the morning of Thursday, the nineteenth, after only a brief rest, James Phelan had breakfast with his sister in their large Victorian house, during which they discussed the probability of organizing the servants to move the household. They were hoping for the best but realized their safety would depend on the firemen's ability to hold their line against the fire at Market and Mission streets, where it had already jumped over

Van Ness. They agreed to inventory the items that must be saved, but because their mansion sat on quite a large piece of property and the chance of the fire's connecting with it was slim, they would wait until absolutely necessary before abandoning their home.

Together they walked to the garden, where they saw a snowstorm of ashes flowing through the sky. Miss Phelan's maid approached them, and Phelan asked if she had heard any news of the Phelan Building, whether it had passed safely through the night. "At that moment," Phelan wrote, "[she] picked up a check at my feet, and handed it to me. It turned out to be a cancelled check, slightly scorched, and about four years old. I at once said, 'The building has been destroyed.' She said, 'How do you know?' I said, 'Here is a messenger from the skies.'

"These cancelled checks were kept in wooden boxes, year by year, and stored in the garret of the Phelan Building. The fire must have burst the boxes, and the contents scattered. The building was about two and one-half miles from the garden, and it has always been regarded as a curious thing that one of these tell-tale checks should have been deposited at my very feet."

Phelan then set out in his automobile to commence his work as head of the finance committee of the Committee of Fifty but only reached as far as 6th Street and Market before being stopped by obstructions from the fire. He realized as he drove through the city that the time had come to evacuate his house, and he returned "to gather up some goods although I was not convinced my residence would be destroyed." He hailed a passing express wagon and arranged for the driver to take two loads for sixty dollars, a fair price. Phelan then began to supervise the loading of "things which Miss Phelan desired, including a Swiss bedroom set which she had purchased abroad, richly carved, wearing apparel, some books, small objects of art, my scrap books and other things." He then organized a minor caravan to take his family, household staff, and possessions to a safe area he had found in Golden Gate Park, where he would camp among clerks, streetcar conductors, and construction laborers, all now one class seeking protection.

Phelan had two automobiles. That day the superintendent of the Spring Valley Water Company, Herman Schussler, came to ask if he could borrow the large Mercedes that Phelan had recently bought new in Paris. Phelan quickly agreed, and the car was used for the next four days to go through the debris-strewn and dynamite-shattered streets to check water cisterns and broken pipes until it was finally towed back to its owner in a substantially reduced condition. Phelan also had a Renault that his driver was especially proud of and cared for as if it were his own. That evening as they were driving, a fireman stepped into the street and announced that the car was being commandeered to pick up a load of dynamite. As his chauffeur urged him to refuse, Phelan showed his badge and informed the fireman that he was, as a member of the Committee of Fifty, immune from such an order. But observing that the need was great, he invited the fireman to ride with him in the tonneau to the Kentucky Street car barns. After loading the dynamite, they headed for 21st Street and Guerrero, where the army would be waiting for the delivery. On the way there Phelan asked "whether there was not a danger of [the dynamite] exploding." The fireman answered that he thought not, but that the dynamite caps he held in his hand could. Phelan later commented, "The care with which my chauffeur [then] drove the car over the rough roads was extraordinary for him. He seemed to be going over velvet streets."

Later that night Phelan returned to his home, and saw that his optimism had been misplaced, for the cypress trees in the front yard had gone up in flames and thrown up many embers that finally took his house.

CHAPTER 68

Battalion Chief John McClusky did everything he could on Thursday morning to keep the fire from crossing Pacific Street. He was relieved to see that Oakland's fire chief, Nick Ball, had arrived with thirty fresh men, but still the paltry supply of water hindered their efforts. Finally, he had his men lay a line of more than four thousand feet from a fireboat just west of the Ferry Building to Broadway and Mason streets, a distance of fourteen blocks. He used three steam pumpers spaced at equal distances to pump the water through the more than forty lengths of hose. The exhausted men tried to stop the fire from going northward across Broadway and reaching the north wharves, where today's marina district is, but it was a vain attempt, for a single line of hose was simply not enough to cool the ever-expanding fire. Captain Henry Schmidt of Engine 8 and his men also had to fight at this location to save their steamer. They had given up their hose, which by now was burning through, and wrapped wet sheets of burlap around themselves as they made a final dash for the steamer. With great effort they were able to pull it far enough away from the flames to hitch it to the horses. As they pulled back north, one street closer to North East Street, they saw the flames jumping across the expanse of Broadway.

That evening a major fire raged in the commercial district. The firemen there believed that they would soon have it under control, but the hose, which came from the tugboat *Pilot,* burst several times due to the great pressure supplied by the tugboat, pressure that was calibrated to fill a hose line that went on for eight or ten blocks. Each time it happened they had to shut the water down and replace the ripped hose. As the blaze came under nearly complete control, it was decided to move the boat from Greenwich to Jackson Street, a distance of seven blocks. This additional distance gave the firemen enough hose line to make certain the blaze would be thoroughly extinguished, for they now had a good, consistent stream being pumped by the *Pilot.* But then, according to Thomas A. Burns, a citizen volunteer, the efforts of the firefighters were undermined by General Funston. The men had been working under Chief Walter Cook, who had performed so heroically in extricating Chief Sullivan and his wife the day before, and as Burns wrote:

> As we were remaking our hose connections, General Funston came along and issued a new set of orders. The gist of these was that the tugs were to get busy saving the wharves and nothing else mattered. So the Pilot was ordered away from the Jackson Street Wharf, and Chief Cook [the SFFD battalion chief who was directing the operation] took his company elsewhere. The fire that we had well under control started up again and burnt the whole district, clear to the water's edge. I sat down on the sidewalk at Jackson and East Street and wept.

The *Pilot* was a private tug stationed at the Mission Street pier, mastered by Charles Love. After leaving the Jackson Street wharf, it was probably sent across the bay by Funston to pick up dynamite. It is likely that only after its return from Pinole Point on this errand was it again used to assist in firefighting efforts. Unfortunately, it would now be assisting in the fight against the regenerated blaze that it had helped bring to a halt earlier and that had since proceeded north and northwest from the Jackson Street wharf, destroying several dozen city blocks in its progress. By the time the *Pilot* was in action again, Chief Cook reported, the conflagration had reached Union Street and destroyed the warehouses on the southern portion of the north waterfront. Fortunately, the *Pilot* proved its effectiveness once again, and under the leadership of Chief Cook, who was joined by the ever-present Lieutenant Freeman and his men, the blaze was finally stopped just beyond Union Street.

The army tugs had been performing invaluable service during the early days of the fire, using their powerful bilge pumps to provide large quantities of water for the firefighters, especially for Lieutenant Freeman along the south and central waterfronts. Thus, a large patch of the city, stretching one half mile or more from the eastern shoreline was saved.

But on the afternoon of Thursday, April 19, General Funston ordered all army tugs—*General McDowell, Slocum, Mifflin,* and *Lieutenant George M. Morris*—away from their firefighting duties on the waterfront and commanded them instead to transport dynamite to the city or to protect Fort Mason from the flames.

Not long afterward General Funston created something of a serious controversy when he ordered that the *Priscilla* be sent to Pinole Point for additional dynamite, removing it, too, from its potential firefighting duties. So few options were available for delivering water to the fire that to remove one of the most powerful water pumps was questionable at the least.

Large sections of the city were undoubtedly sacrificed to the advancing flames due to the lack of hose streams in the north waterfront. Ironically, many privately owned merchant ships were anchored in San Francisco Bay, and each one would have been capable of carrying the same amount of dynamite that was ultimately transported by all the tugs that were sent to Pinole Point. General Funston could have asked, as Lieutenant Freeman had, any one of these vessels' captains for assistance or he might even have (illegally) commandeered one of the ships.

CHAPTER 69

Henry Lai did not sleep at all on Wednesday night. He walked, sometimes ran, from one street to another as the fleeing Chinese poured from their overcrowded apartment buildings. His throat was parched from the smoke, but he stopped frequently to ask if anyone knew Yuen Kum. All day Thursday, now most often running, he searched the City of San Francisco, realizing how quickly the fire had spread through every last building in Chinatown. He decided to go to Golden Gate Park, for he heard people in the area of Sacramento Street say that many of the Chinese had headed there. He continued to mingle with the crowds, asking in Chinese if anyone had seen Yuen Kum or Mrs. Cameron, but they were strange names, and people did not recognize them. Henry was an emotional young man and felt as bereft as a child who had been separated from its mother, and he could hardly keep the tears out of his eyes. Yuen Kum was the dream of a better life for him, an image of kindness and beauty, and now she was gone. He saw many adolescent girls as well as stooped-over matrons carrying the young on their backs in slings, with suitcases and bags in their hands, and he could not help wondering what his fiancée must be suffering, what she must be going through. She had to be alive. She could not have . . .

Refugees huddled on the roadside, solidifying their plans and checking their packages before they joined the slow march on the road to Golden Gate Park. But there were just as many evacuees going north to Union and Chestnut, through the choking mist to a safe haven by the water, perhaps to find a boat of any kind to take them to Sausalito, Tiburon, or Petaluma. Most of the Chinese, though, headed south on the peninsula toward San Mateo, knowing that they would not have to depend on the kindness of a sea captain to take them across water.

On the road, Henry met people who told him not to go to the Presidio, for the Chinese and the Japanese had been turned back by the military. A baseless rumor had spread wildly among them that Asians would receive no provisions at the army camp, and so Henry decided to make his way to the seaport.

CHAPTER 70

Midshipman Pond wrote his own full account of the suddenly liberated chickens, which he saw as something of comic relief in the midst of anguish. "There were about 150 freight cars," he recalled, "on the spur tracks of the Belt Line Railroad . . . filled with produce, and among them four or five cars of live chickens. Lieutenant Freeman received word from one of the railroad officials to liberate these chickens and turn them over to the crowd, as they would soon die. . . . The ensuing scramble for chickens created quite a bit of excitement, and . . . put everybody in a good humor." But Pond went on to explain, "The precedent set by this unauthorized . . . distribution of food supplies was not so good, however, for soon some men were observed to be tampering with the locks of other produce cars. That some of them were not shot as looters by our armed patrols, speaks well for the latter's restraint and good common sense."

The produce cars in the railyards were to prove vital in feeding people throughout the next two days, before sufficient supplies were able to reach the city. One ship captain was said to have looted foodstuffs, which he stored in the hold of his ship, the *Hartfield.* Freeman, in between the many duties he undertook, paid the vessel a visit and found that not only had the captain, named Sanderson, not stolen anything, but was amenable to having his ship used as a temporary refuge for the city's men, women, and children. Furthermore, the captain and his men set immediately about helping Freeman and the sailors with fighting the fires. Freeman took them on enthusiastically because every hand was needed and asked them to guard the Southern Pacific cars, since this food would undoubtedly be critically important.

On Thursday afternoon the wind again changed and for the first time built up to significant power. The very same streets that had suffered the fire the day before were again now threatened. Buildings that were still smoldering began to leap again into flame as if they had been fanned by a giant bellows. Freeman determined to stop this fire once and for all, and he directed his men to congregate at the *Leslie*. He had them stretch a long line of hose, and when it ran out he sought the assistance of the firemen who brought them more. Ultimately, the hose stretched for more than eleven long blocks, across the spur tracks and up over the side of Telegraph Hill to Broadway. When they again reached Montgomery, for the second time in two days, they positioned themselves firmly in front of the advancing fire.

They did not realize at that time that the army had decided to dynamite the Monkey Block building or that Dr. Emma Sutro Merritt had arrived just in time to beg them to spare it and the more than one hundred thousand of her family's books that were housed within it. After some consideration, the army consented to the pleas of this prominent woman, and once the powder was removed from the building, Jack Murray, his captain, and the men of Engine 1 were ordered to protect it from the interior. They worked for the next several hours in the great heat and smoke, smothering the fires as they started at every window frame. When the fire had again safely passed, the men of Engine 1 returned to the fire line to join Lieutenant Freeman and the sailors in the continuing fight.

They not only saved the Monkey Block, but again stopped the fire from crossing over to the U.S. customs house. They once again saved the Hotaling Whiskey warehouse.

It was always a problem to find hose that connected easily because the coupling sizes and the thread count used by the navy were different from those used by the fire department. But the men managed to find five hundred feet of additional compatible hose, which enabled them to go down Montgomery to New Montgomery, which is today, after the large migration of Italians to San Francisco, appropriately named Columbus. While waiting for the water to come up from the *Leslie,* Freeman, according to Pond, was fierce in his determination to keep the fire from crossing Montgomery again. He led his men, sailors and firemen both, "through large buildings, and tore down all inflammable material, such as awnings, curtains, etc., and climbed to the tops of the buildings to beat out the fires started in the cornices and window frames by the terrific heat from the blazing mass across the street." All the while Freeman remained in front of his forces, watching for their safety and guiding their tactics. "A born leader of men," Pond wrote, "I can hear him now, 'Come on, men, sock it to 'em.' And, they did." In a big wooden hotel they went from room to room taking whatever liquid they could find in any jar or receptacle and "dashing it on the burning cornices."

Midshipman Pond remembered that the San Francisco Fire Department had small and easily transportable chemical wagons, which though useless against the bigger fires might surely serve in dashing the little flare-ups. He ran to Chief Murphy, who was still at the scene, and asked for a chemical engine. Murphy found one that was immediately put to use and quite effectively. Because the firemen had to operate the chemical apparatus at the front of the buildings, though, they were forced to deal with the raging conflagration at their backs. But they held their places in searing heat. Someone had found sheets of canvas, which they kept wetting and used to cover the firemen's shoulders. But the heat was still so intense that the men searched for whatever puddles they could find, rolling in the water and mud, completely saturating themselves. As they took their places at the fire line, however, they dried off completely in almost no time at all and were again being singed. In all probability, Jack Murray was in this group of firemen and sailors, for Chief McClusky reported that he had sent Engine 1 to Mason Street, and this area was in Engine 1's first-due district.

Jack also saw that the fire had begun to do what he had been praying it wouldn't—making its way up Telegraph Hill. He knew his family would have to evacuate, but even though they were only a few streets away, he could not take leave of his fellows on the fire line to help them. He thought of young Will, and how impossible it would be for Annie to persuade his son to leave his beloved fire engine behind, and how equally impossible it would be to tow the big toy fire truck along, to find room for it in an evacuation on a carriage or a boat. Fortunately, he spotted a fireman he knew, an operator for the department's messenger and equipment delivery system, sitting in the bucket seat of a service cart pulled by a horse. Perhaps he would help.

CHAPTER 71

On Thursday afternoon, H. M. Griffith and his family stood at the window of their living room, astonished at what lay before them. With his camera at the ready, H.M. took another photographic plate every time the fire advanced up Montgomery. To the family's amusement, they watched a woman hang clothes on the roof of a building immediately below them, going about her business confident that the firemen would have it all under control soon.

The Griffiths grew anxious, though, as the fire began to come up the Washington Street hill. Until then, they had prayed—just as the members of the Bohemian Club, the owners of the Fairmont Hotel, and the officials of the Hopkins Institute of Art had prayed—that they would escape the onslaught, but prudently they had already packed those family possessions that had to be saved. They had just a few trunks, and the Chinese houseman tied as much together as he could and began to carry them down the stairs to the street. Mrs. Griffith dressed in her finest clothes, for there wasn't room for them among the family heirlooms, letters, paintings, daguerreotypes, and photographs. Mr. Griffith took one last photograph from the window and then folded the tripod and followed the houseman. Before leaving, Mrs. Griffith, strong and unsentimental, took a final look around her home, rummaged through a closet for an appropriate hat, and adjusted it carefully on her head. She shooed her son out of the apartment and, remembering the experiences the family had shared in these rooms, said a silent prayer that they would one day safely return.

In the street Mrs. Griffith sat on the small amount of the family wealth, on the sidewalk next to the large water reservoir of the Spring Valley Water Company. The tank itself, which sat on top of a large, grassy knoll, was empty because the water had flowed down the hill when the main in the business section below shifted in the earthquake and broke apart. The houseman sat behind her as they waited for the men of the family to find a carriage to take them to a ferry. The wind was pushing their way, but still no one was choking or had to tie handkerchiefs around their mouths. Most people in the neighborhood were calm and set about putting some plan into action. The Griffiths had been told about a refugee center in Oakland and that is where they were determined to go. In the photograph taken of Mrs. Griffith on Washington Street, it seems she had just one momentary hope on her mind—that her husband would find a way to take them to safety.

CHAPTER 72

Henry Lai was beside himself with frustration and annoyance because the soldiers kept pushing him with the sides of their rifles every time he tried to cross the street into a burning Chinatown. "But my wife is there," he must have kept saying, pointing a finger at the fire. In his heart he must have believed they were already married, for the vow and the promise had been made, and it was only incidental that Miss Cameron had arranged the big wedding so that God would approve and bless their union. Henry began to think of all his happy plans, and now what was he to do but cry as he kept begging, "Please let me by." But the soldier had his job to do; as far as he was concerned, this was just another Chinese whose cause was unimportant.

At the same time, unknown to Henry, Donaldina Cameron was shepherding her fifty children through the street only a few blocks away. A poet might think of them as a small army of angels, each holding the hand of the one in front and the one in back. "Just keep going," she yelled when the rhythm of their escape slowed or the line of the little ones wiggled, "and don't stop."

"Fire menaced from three directions," Donaldina wrote. "What tragedy . . . crowded into our lives these two days. . . . Never shall we forget the hasty preparations made that Thursday morning for the long march to the ferry. Many things carried so far must be left behind: much must be carried. . . . All had a load, not even little five-year-old Hung Mooie was exempt. She tearfully consented to carry two dozen eggs in the hope of having some to eat in the by and by. . . . As tears would not avail (the hour for weeping had not yet come), laughter was the tonic which stimulated that weary, unwashed, and uncombed procession on the long tramp through stifling, crowded streets near where the fire raged, and through the desolate district already burned, where the fires of yesterday still smouldered."

Donaldina had sent word ahead to Sausalito, and she was hoping that her messenger had arrived and preparations being made for the large crowd she was bringing. She and her charges were headed across the bay to the seminary at San Anselmo, to the ivy-covered buildings overseen by Dr. Landon, one of the warmest, most generous of ministers she had ever met. She prayed as she stroked the heads of the girls, many of them running to keep up. God would help them, and they would be received at San Anselmo. But God had so many other supplications in that great mess of a day, and what would she do if He could not find time to answer her one request?

CHAPTER 73

It was not like Annie Murray to wring her hands, but she was at her wit's end deciding what she was going to leave behind to be sacrificed in the fire. She knew the order to evacuate would come at any moment, and she felt as if she were being forced to pick out little pieces of their lives from room to room and to abandon other pieces of their lives forever. She pulled a trunk out from a high shelf in her closet and began to pack it with photographs, the fine linens from her wedding, the silver service that she had collected piece by piece, a set of antique cups—anything small that was wrapped in memory. She put her pots and pans, sheets and pillowcases, and the children's clothing in navy duffel bags. She reached for the pillows but then left them—pillows could always be bought. The children were each doing something to clean, fold, or pack, and the family was fully occupied when the heavy knock came at the door.

As Annie opened it she saw the Italian family just across the street, all six of them, in the street with mops, rags, and several buckets of red wine, washing down the side of their house. They were certain to save it by keeping it damp, if even with homemade wine. Annie wasn't sure if she recognized the fireman at the door, but she knew the uniform and invited him in. "No, ma'am," the fireman said. "I just came to have a talk with Will."

Will had been standing right behind his mother's long thick skirt, and Annie took his hand and drew him forward as gracefully as in a dance movement.

The fireman turned his attention to the boy, smiled, and began an earnest conversation with him. "Will," he said, "I have come to ask of you a mighty favor, and I hope you can help us out. You know the fire is moving pretty quickly through the streets, and it is coming up to our firehouse. We have to move everything out, and we need to find everything we can on wheels. We know you have that big fire truck, and we could put a lot of things on top of it if we had it."

Will suddenly looked frightened. He saw the horse and the service cart in the street, and he knew the fireman had come for the best present he had ever been given. He thought of how he and his father played a game in which they blamed an invented Phil Priggens for every misadventure, and he wondered for a moment if Priggens had sent this man. But then he realized it could have only been his father who told this fireman about his fire truck.

"It will surely help the fire department," the fireman repeated, and that convinced Will that he had to surrender the toy. It was right behind him, and he simply bent down and gently pushed it forward. The fireman picked it up—it was so big he could barely get his arms around it—and carried it to the service cart, stowing it sideways so that it wouldn't roll off. He turned then, gave a short wave, and drove off down into the smoke-filled streets below.

Will held his head high as he waved in return, his mother's hands firm on his shoulders. The Murrays would not have Annie in their lives for very much longer. She would develop an embolism in her brain and die prematurely only six years later, and would not live long enough to see Will follow in his father's footsteps and raise his hand for the oath of office of a San Francisco fireman. But she could not have been prouder that day than to be part of a fireman's family.

CHAPTER 74

The Bohemian Club had been founded by a group of prominent San Franciscans in 1872, and of all the San Francisco clubs it was the most serious minded. It was created so that gentlemen—no more than 750 of them—who were connected in some way to literature, art, music, or drama could come together in mutual interest and satisfaction. It so happened that James D. Phelan was the president of the club in 1906 (though he does not mention the club in his account of the fire). On that fateful Wednesday morning a group of these Bohemians had rushed to the club building, then at 130 Post Street, and, assessing the damage to their beloved organization, debated whether or not to immediately remove all of the important artwork and first editions to a safer place. They owned a significant collection of nineteenth-century American and European paintings and sculpture, all donated by the well-heeled and well-traveled members. "But optimism asserted itself in confident tones," stated the annual report of the club, "and [we] began to predict that the dynamite squads surely would not permit the flames to cross Market Street." Like many in the city, the Bohemian members believed that the fire could not grow out of the control of the fire department for, after all, the army was in full force backing it up.

That their confidence in the army's dynamiting efforts was misplaced became evident on Thursday afternoon, when the fire began to sweep up Post Street. Charles Dickman, head of the art committee, and with the help of Henry the hallboy and others, began to pry the paintings from their frames, racing against the approaching flames. A truck from the Vulcan Iron Works was stopped and hired, and as it was carrying a full load, members and others worked together to remove its cargo. The art was then placed in the truck, each piece wrapped in a blanket, sheet, towel, or even the old jackets that had been left in the club for members to wear if they arrived without appropriate attire.

All must have been proud of their salvage, but as soon as the truck was completely packed, a squadron of soldiers appeared and without much ceremony confiscated the vehicle for their own use. There was little argument, for everyone believed they were subject to the general rules of martial law, and the members took the paintings back into the club until they could decide what to do next. As they debated, the fire made its inexorable progress up the street, building by building.

Just then, one of the members, James McNab, appeared like a deus ex machina with two wagons to cart away the club's patrimony to a tented refuge in Golden Gate Park. The rescuers also retrieved the club's financial records, and afterward some of the members, never willing to be without controversy of some sort in their lives, accused the art heroes of lightheadedness in also saving the individual due accounts of San Francisco's leading citizens.

By nightfall of the second day the clubrooms were gone in a blanket of smoke, but the heritage and legacy of the Bohemians were, notwithstanding soldiers or flames, safe in the park.

Just a few blocks away, on the corner of Nob Hill (and now the site of the Mark Hopkins Hotel), stood the solid and palatial former residence of Mark Hopkins, which had cost him about two million dollars to construct. It was now owned by the University of California and housed the Hopkins Institute of Art, a school of nine teachers and about two hundred students. The building housed several hundred pieces of art, both paintings and sculpture, which were on exhibit to the general public. The men and women who were in the school that Thursday became frantic as the fire came up the hill. Helped by an army officer named Lieutenant McMillan, the students and faculty desperately tore the paintings from the walls, working in smoke, coughing and suffering fits of dizziness. They regretted not having removed the art earlier, but like the Bohemians they all believed the fire would be stopped long before it arrived at their front steps.

They stored some of the pictures in the basement of the Leland Stanford mansion, confident that, with its sitting at the very peak of the hill, it would survive any onslaught. But the paintings were lost in the complete destruction of Governor Stanford's home. The only artwork from the institute that was saved was a group of about ninety paintings that had been carried home by the students and teachers who lived a safe distance from the flames.

In all there was a registered loss for the Hopkins Institute of $573,000, and with insurance coverage of only $143,000 the school faced total ruination. But in its annual report, the board of trustees declared its commitment to rebuilding, and like the rest of San Francisco, they tried to look only to the future.

CHAPTER 75

Late Thursday afternoon, Annie organized the few bags and duffels that she and the children had carried down the hill. They were at the wharf, at the edge of Pier 7, in a gathering crowd of refugees at a spot where they were told to wait to be taken away. After making an orderly pile of their belongings Annie turned a trunk on its end so that she could sit as they waited for a schooner, a ferry, or any boat at all to carry them to Oakland. Suddenly, Annie rose to her feet, realizing that in the frenzied rush to leave their home she had forgotten the most important package of all, the trunk with all the family memorabilia that held her wedding presents and silverware. She became exasperated and frantic, and was even more disconcerted than she had been leaving her home, for its loss underscored for her the comprehension of how their lives were so unalterably changing.

"Quick, Josie," she cried to her oldest, "run up the hill and get the old trunk by the back closet. Just drag it down here any way you can." That trunk held the valuable few photographs and daguerreotypes from both the Murray and Jordan families, and it would be so hard to secure the collective family memory, to tell the stories of their life in the West without them.

Josie pulled her skirt to her knees and ran—all the Murray children, like most children in San Francisco, could tirelessly climb hills. Annie paced the cobblestoned ground of the pier for the next half hour until she saw Josie in the distance. Just behind her on the bay, a boat was coming in to the landing, and beyond Josie, she could see that the smoke and flames were headed directly for Telegraph Hill. So she began to pray as

Josie got closer to them, struggling as a fourteen-year-old would with a large trunk, pulling it by one large leather handle. Annie was the kind of woman who would say a prayer rather than let life's disappointments get the best of her, and she now began to pray because she saw immediately that Josie had brought the wrong trunk—the one that was *in* the closet instead of the one *by* the closet. The trunk Josie was so determinedly hauling held all of Annie's sewing scraps and old dresses and slips, material to be used for repair and adornment, like silk garnish on the few things she made for the girls. She hid her disappointment; Josie had done her best, and the house would soon be gone. The boat had docked, and people were lining up. Without words, Annie smiled as her oldest child carried the trunk aboard.

CHAPTER 76

Through Thursday night and into Friday morning, the fire swept over Telegraph Hill and traveled north and west to Russian Hill. There were many reports of how Jack and Annie Murray's neighbors on the top of Telegraph Hill had saved their homes, as had the residents at the top of Russian Hill, a solidly middle-class enclave of people trying to keep their lives together. There were dozens of homes here, residences of respected San Franciscans like the Reverend Joseph Worcester, Livingston Jenks, and Eli Shepard. There were groups of three-story flats and other properties owned by a Mrs. Williams, a Mr. Livermore, and others.

In one report, by Mr. Laffler, it was suggested that the homes at the top of Russian Hill were saved, for the most part, in spite of the military. The buildings caught fire several times, and each time the soldiers demanded that the hill residents evacuate. Yet the Russian Hill families stubbornly resisted each effort to remove them, and they continued trying to extinguish the flames by beating at them with rugs, brooms, and little pails of water and wine. It was reported that one home was saved completely from the flames by a barrel of vinegar that was being made in the cellar. At about 8:00 on Thursday evening, the soldiers trooped up the hill, determined to oust the families who were willing to risk all to save all. The soldiers went through the homes with fixed bayonets, threatening every human being they found, forcing them out of the house.

With every such account it must be asked again if much of the city could have been saved if its citizens, resolute Westerners with much courage, had been allowed to put themselves at risk attempting to save their own property. And what would the consequence have been had entire neighborhoods been allowed to group and mobilize themselves? Yet in posing such questions we must not forget that every citizen in the city believed that martial law was in place, with a legitimate shoot-to-kill order. Funston and Schmitz—and to the degree that they could not see viable alternatives, the Committee of Fifty—forced the population into a powerless position. Their only choice was to outwit, outrun, or circumnavigate a small army of soldiers.

Tragically, in some areas where the local residents had been successful in defeating the blaze, the army would promptly undo their work by forcing the evacuation of the area. A gruesome example took place farther down Green Street, just east of Van Ness, where troops dynamited an area just saved by twenty local citizens.

An observer, Porter Garnett, wrote:

> I was watching the fire with special reference to a friend's house on the north side of Green Street near Larkin and had concluded that it was safe. No fire was visible north of Green Street, and on the south side of Green Street the flames appeared to have been completely extinguished. A few moments later I again looked from the window of the house in which I was on Pacific Avenue—a house commanding an excellent view of the district in question—and was astounded to perceive several isolated fires in the district which a short time before had seemed to be free from danger. These blazes were quickly fanned by the wind into a roaring conflagration, and the house of my friend was within a short time burned to the ground.

CHAPTER 77

The firefighters were nowhere to be found on Telegraph Hill as they had concentrated themselves at Montgomery, up Nob Hill, at Van Ness, and in pockets at the Mission district. The fire department had finally determined to make a stand at Van Ness because of the width of the boulevard, and its men stretched a hose line from the bay, up Van Ness, to Sacramento, a distance of more than a mile, perhaps the longest hose stretch in firefighting history. Three steam pumpers were used to relay the water. In this, one of the most heroic and unheralded events of the fire, the firemen had to call on every last ounce of strength to carry and pull so much hose for so long a way.

On Friday morning, however, the decision was made to dynamite the entire east side of Van Ness, from Union to Geary. The army first brought in its heavy artillery and began to shoot heavy shells at the houses, hoping to weaken the structures, but they had just limited success, for the houses, all expensive mansions and apartment hotels, were in most cases large and well built.

The artillery was also proving dangerous, because it was hard to control the consequence of exploding shells, and the firing was soon stopped. The fire department crew and Lieutenant Briggs, using whatever dynamite and black powder they could find, now began to dynamite the buildings on the east side of Van Ness, at Geary. But the wind had been coming from the bay at the time, and it would have been better to dynamite into the wind on the lower end of Van Ness, near the city hall area, for fires were created with the first blasts that made it impossible to get into the adjoining buildings. Again, instead of stopping the fires, the dynamiting simply started new ones that were hard to control and that, indeed, jumped across Van Ness and burned up four blocks on the west side of Van Ness, from Sutter to Sacramento. A grueling pessimism must have overtaken General Funston at this time because after watching how the fire had progressed he wired Secretary Taft at the war department:

ALMOST CERTAIN NOW THAT ENTIRE CITY
WILL BE DESTROYED.

But so dire a prediction was unfounded, as the city's firefighters were beginning to discover. As Chief Shaughnessy saw that the whole of the Western Addition section was threatened with destruction, he employed his men quickly and forcefully to stop the fire at Franklin, just one block west of Van Ness. It was a tough fight, and it would get tougher as the day wore on, but, then, an anomaly occurred, one that no firefighter could possibly have predicted.

It was within this four-block area on the west side of Van Ness that Claus Spreckels's sixty-room mansion was located. The Spreckels family and staff had already been forced to vacate by the soldiers. The building sat on the south corner of Clay, and it was the most majestic of all the important buildings—mansions, hospitals, government buildings—on Van Ness. Claus's neighbor, cable car president James B. Stetson, himself the owner of one of the most photographed mansions in the city, had circumnavigated the military guards and watched as his neighbor's house, "one of the most beautiful buildings in the country, with its solid silver doorknobs and bath fixtures," burned to the ground. The fire had crept up to it from the rear when a fence and a garage caught fire. It was at first a small fire, and Stetson asked some soldiers and firemen to extinguish it. They paid him no mind, and Stetson reported that one bucketful of water would have been enough to save the whole house, for it was the last to burn on that side of Van Ness. In fact, the fire was stopped there, and the buildings on the north side of Clay remained unscathed. It is a mystery why, considering Stetson's report, no effort was made to save the property, particularly since the firemen had stopped the fire at Franklin and had certainly been in the area. Perhaps the firemen were legitimately preoccupied, or had, in the continuing crisis, been ordered to another street? Or as some have suggested, did some political leader with a grudge against the Spreckelses order them away? From the beginning it has been speculated—without any evidence—within the pages of important books that Abe Ruef might have paid an arsonist to burn down Rudolph Spreckels's mansion, for he had correctly assumed that Rudolph had important evidence in the corruption case stored in his home. The arsonist, however, confused the mansions and burned down Claus's by mistake. The truth will never be known without the discovery of the long-lost Committee on History material, unless it is found in some document that is safe, yet forgotten, in a family's attic.

CHAPTER 78

During this time refugees wandered the city at the fringes of the fire, searching for places to homestead and to deposit their scanty belongings, whether in graveyards, empty lots, or plaza parks. Loretta Cutting Underwood had been living with her family and a few boarders at 513 Fulton Street, a thoroughfare that then ended at the steps of city hall, on the triangular property at Larkin and McAllister. In the first minute of the earthquake Loretta saw the flue above her stove dislodged so that the heat of the stove's fire openly licked at her ceiling. But then, she reported, she watched the work of Providence, which knocked over her large, open-topped kettle, pouring water conveniently over the stove fire, extinguishing it enough to keep it from igniting the ceiling. By the afternoon, the conflagration had come to within two blocks of her home, and since there was no one to hire to move her household, Loretta realized she had to find some protection nearby, for the closer to home she could locate an open space the more she would be able to cart to safety. She came upon a lot on McAllister where about a dozen families were already camped out, many staking their claim to space with improvised fencing.

Loretta wrote:

> The little alley bounding our lot on the south was strewn with all sorts of household belongings. People would come along, see an easy chair, and drop down into it exhausted. Another would come and wheel a trunk or dresser up the street to some other place where they had camped. Others would race in with their belongings, get scared at the fast approach of the fire, then race out again. Some would be hauling their

> things through the middle of the street in one direction; others would be hauling theirs as madly in another. Someone would saunter along without any belongings, apparently [with] nothing to do, no place to go, [and] would open one of the many pianos standing in the street, play a few bars, a crowd would gather round, some well-remembered song or hymn would be started, the crowd would join in. Then perhaps in the middle of a sentence the earth would begin to heave and subside, and then repeat with a few side waverings; and the melody would break, the words would cease, and finally the music would become inaudible, the piano would be closed, and the crowd would drift on . . . an endless chain . . . as they march by.

Many aftershocks of diminishing strength occurred through the first week of the earthquake, and each one brought with it a dreadful, if unfulfilled, anticipation of another catastrophe. Fortunately for Loretta, her home was just on the west side of Van Ness, and so she was allowed, under escort of a soldier who would ascertain her ownership, to return to it in a few days' time.

CHAPTER 79

By midday on Thursday, hope for rescuing San Francisco was placed in saving the areas of the southern part of the Mission and Potrero districts east of Dolores Street and below Mission Dolores, and also the hundreds of homes in the Western Addition, west of Van Ness. Chiefs Dougherty and Shaughnessy directed their men to each of these two neighborhoods and in many cases had to plead with the army officers to let civilians help with the firefighting in the evacuated areas.

A writer named Charles Keeler wrote that automobiles continued to pass on all the avenues carrying finely dressed women with Indian baskets on their laps filled with bric-a-brac and little treasures. He also saw two men who rolled a piano down Van Ness until its casters crumbled, and then in their determination to save it began to push it end over end as they hurried down the street, as if it were a crate.

By mid- to late-afternoon Thursday, H. M. Griffith and his family were on a boat to Oakland, from where they saw that San Francisco was enveloped in so much smoke that it was hardly visible to the outside world. Even under these conditions, Mr. Griffith could not resist snapping a photograph of the vague image of the Ferry Building in the distance. When they reached Oakland, the Griffiths were taken to a refuge center, perhaps organized by a church or civic organization, and were given an evening tea at a properly set table in the back storeroom of a store amid boxes of pork and beans and canned tomatoes, a far cry from the elegance they were accustomed to before their home at the top of Nob Hill burned to the ground.

Egbert H. Gould, a Chicago businessman who was staying at the Palace Hotel when the quake struck, wrote an account of what he saw from Oakland at about 5:00 P.M.: "At that time, San Francisco was hidden in a pall of smoke. The sun shone brightly upon it without any seeming penetration. Flames at times cleft the darkness. This cloud was five miles in height and at its top changed into a milk white."

At about 7:00 P.M., the principal ships of the Pacific Fleet, under the command of Admiral Caspar Goodrich, finally reached San Francisco, having traveled at full steam since the afternoon of the previous day from the seas just off the American and Mexican border, near the City of San Diego. The slower ships of the fleet continued to enter the bay over the next few hours, bringing supplies, water, and manpower. The seamen would work with Lieutenant Freeman, the medical centers, and the relief camps, and provide much assistance to the city in the next days and weeks.

Since Lieutenant Freeman had sent the *Sotoyomo* to Goat Island for another cargo of water, it had been unable to return in time to help the firefighting efforts along Townsend Street. The *Leslie* alone was left to supply pumped seawater for the firefighters and the sailors, and Freeman must have been reassured in his decision to keep back some of the tugs. The fire they had previously stopped at Townsend began again to threaten the area just to the south, where the rail lines run to the sheds, just as the wind started to carry the flames through the North Beach section and up Telegraph and Russian hills. Freeman, who was fast learning the fireman's constant dread of the one sparkling ember left in a wallspace that can create a major rekindle, directed his men to run the hose to Eighth Street, and they fought the blaze there for several hours, finally stopping it, between Brannan and Bryant, by the early morning on Friday.

By this time the troops from the Presidio had moved their patrols to the inland part of the city and left the entire waterfront for a few navy personnel to guard. Freeman had assigned marine First Lieutenant Smith from the *Active* and his five-man squad to the task. Lieutenant Smith and his men were able to stop any looting, closed the few remaining saloons still in operation, and assisted with the relief efforts for the refugees. General Funston had provided Freeman with an additional guard of marines to support the patrol of the several miles of wharves, piers, and business establishments, and with these forces in hand, the waterfront was soon under control. However, sometime in the afternoon, Funston relocated many of the marines to the United States Sub Treasury on Commercial Street. Lieutenant Sidney Brewster took about twenty men for this guard duty.

Medical supplies begin to arrive for the harbor emergency hospital from Marin, San Mateo, and other neighboring counties, but the most urgently needed commodity was freshwater, as the people continued to suffer from parching thirst. Freeman had a conversation with a Captain G. B. Musson, master of the British merchant vessel *Henley,* who then agreed to start his evaporators and begin to distill water. Captain Musson kept his evaporators running for several days, performing an invaluable service as the need for freshwater on the waterfront remained constant.

Late Thursday night, as the fire crossed a five-block-long section of Vallejo Street, no group of firefighters of any kind was available to stop it. Though by now there was a reported glut of unofficial law enforcement officers, special police, railroad men in uniform, and self-appointed law enforcers, there was a desperate shortage of firefighters, and no attempt was made to make a stand at Vallejo.

As Freeman had believed, if only a few hundred extra men had been on hand to halt the progress of the flames, they might have been able to save the entire area around Russian Hill and the north waterfront, an area of one hundred more blocks, every one of which was destroyed. Even here, however, desperate families valiantly struggled to save their property. William Keller and his sons were willing to give their lives for the business they had worked so hard to grow into one of the most important mills in the West and to preserve their building and the ten thousand barrels of flour and forty-five hundred tons of wheat it contained. The Globe Mills was located at the base of Telegraph Hill, just one street from the waterfront, and the Kellers felt as ready as anyone could be in the face of an encroaching fire, for they had two hose lines, a water pump, twelve fire extinguishers, an unlimited supply of water that came directly from the bay through a piping system, brick construction, tin-cased windows, a metal roof, and the experience of plenty of fire drills. The mill doors were made of iron, and the mill itself was not attached to any other building. In fact, William Keller was so well prepared to deal with any possible fires that he considered the idea of spending hard-earned money on fire insurance to be a frivolous and needless expenditure. What the Kellers had not counted on, though, was the presence of the army.

On Friday afternoon, as the fire was approaching from a burning lumberyard to the west of Chestnut and Montgomery, where Globe was located, the Kellers gathered ten of their men and positioned themselves in wait, confident that they could stop the fire at the door. But at 4:00 P.M., the soldiers came and gave only one command: "Out!" The Kellers protested, and shouts filled the warehouse. Keller reported that when he insisted he had a right to die in protecting his own property, a soldier challenged, "If I wanted to die he would oblige me right there and now—by shooting me and my men on the spot. If we wanted to live we had to get the hell out of the place." The soldiers then began to prod the Kellers and their employees with the points of their bayonets. Continuing their protest the men of the Globe Mills were forced to leave their premises.

Before he moved permanently to Los Angeles, Mr. Keller testified in a hearing: "We left the building, and late at night, after being exposed for many hours to the heat of burning lumber yards to the north and east, windows to the east front at length broke, and bins of wheat thus directly exposed to the heat, were ignited. There is of course no doubt whatever that one man could have saved the structure had he been permitted to remain."

CHAPTER 80

All fires are eventually extinguished or burn out of their own accord. By early afternoon on Friday, there were reasons to expect that the great conflagration of San Francisco was nearing its end. The political situation in the city was deteriorating even as the fires began to come in check. Early in the day at a meeting of the Committee of Fifty, Mayor Schmitz called for a motion to bring even more armed patrolmen into the city. As one writer recalled, "James Phelan and Rudolph Spreckels opposed the suggestion of another armed force in the city," and the mayor became enraged, berating and belittling the committee for its lack of support during an emergency. The committee, stunned, capitulated, and a protection committee of two hundred men was approved to be under the direction of a Mr. Julian Sonntag. Sonntag then proceeded directly to Fort Mason to gather guns and ammunition for his small private army, but he was turned down. This further infuriated the mayor and provoked another outburst at a meeting of the Committee of Fifty.

In the meantime, the firemen were standing firm against the fire at Van Ness, and for the first time they were not being pushed back. By midday Friday, the fire in North Beach had gone over Telegraph Hill, where it leveled Jack Murray's home, and over Russian Hill, where many houses were saved by the community, and had taken just about everything to the north waterfront, near where Fisherman's Wharf stands today. It had burned itself out as it neared Meigg's Wharf, where as many as thirty thousand people had gathered, waiting to be evacuated. The firefighters, ever aware of the possibility of a rekindle, felt that for the most part the fire in the north waterfront was then under control. The only remaining major confrontations were at Van Ness and in the Mission district, where the fire was still moving from house to house, street to street.

Reassured, Mayor Schmitz sent a telegram to President Roosevelt:

WE ARE DETERMINED TO RESTORE TO THE NATION ITS CHIEF PORT ON THE PACIFIC.

So within a matter of hours Washington officials received two contradictory messages sent by two of the highest ranking city leaders—Schmitz's promise of impending salvation and Funston's warning of impending doom.

Even as an end to the fire did seem to be in sight, for reasons that have never been adequately explained, the soldiers began again to shoot their artillery at the Van Ness houses and to set dynamite. One of the buildings that they attempted to blow up was the Viavi Building. "On Friday afternoon," reported Robert Royce, an attorney then with the firm of Lewis & Royce, "I witnessed the destruction of sundry buildings on the east side of Van Ness Avenue, the avowed purpose of the dynamiters being to prevent the fire from crossing that street. The men seemed to be indifferent regarding the spread of the fire to the east over the unburned section north of Green Street. I saw the destruction of the Viavi Building by dynamite, and I have no doubt that, as frequently asserted, brands hurled by this explosion set fire to this section and were the cause of its destruction."

The fire caused by the dynamiting of the Viavi Building would go on to consume fifty square blocks of San Francisco. It would burn through the day and night, from Green Street and Van Ness to Filbert, then to Polk and to Greenwich, where it turned down Larkin and headed to Jones. It then went to Bay Street, where it turned on Taylor and headed for the waterfront and Lieutenant Freeman.

Mayor Schmitz was reported to have been visibly upset by the dynamiting of the Viavi and to have subsequently had an argument with General Funston. He informed the general that the dynamiting must be halted forthwith, and Funston is said to have left the scene in great anger. He then reported to Fort Mason, where the fire was threatening the army installation but never extended to it.

The new fire caused by the dynamiting of the Viavi would demoralize and cause the collapse of many of the already-spent firefighters. One writer would report, "Dougherty's men had fought their hearts out and all they had to show for it were a few buildings—islands in a sea of ashes. The fight was gone from them." Doctors and nurses who had volunteered to be on the scene also noticed that the firemen were near prostration from exertion and lack of sleep. One of the city's physicians thought that they could be tendered a boost of needed energy if they were given an injection of strychnine, and so doctors and nurses went up and down the fire line on Van Ness administering the needles to the weary firefighters. It was a dangerous and unhealthy remedy, but it did seem to work, and the firefighters found energy on the fire lines they never thought they had.

Neither Jack Murray nor any of the hundreds of the city's firemen abandoned their posts during this long fight. They had been toiling for almost three full days without more than ten or fifteen minutes of rest here and there. They had little idea of the condition of their homes or the safety of their families. More than half of San Francisco's firemen lost their domiciles to the fire. But, still, they stayed the fight.

Because the Viavi fires were headed to the north toward the waterfront, where Jack and a large contingent were already posted, the firefighters at Van Ness turned their attention to the Mission district. Chiefs Shaughnessy and Dougherty led their men south, sometimes struggling to get their hose carts and steamers up and down the slopes.

As was the case in South of Market, much of the Mission district was founded on alluvial soil, and the earthquake cracked many streets or caused them to sink four or five feet into the earth, upending trolley tracks, in some places as much as five feet. Fires were raging through many areas in this part of the city, and in one place a blaze was four blocks wide, generating an enormous amount of radiating heat. To counter it, firemen tore off the side entrance doors to houses and held them like shields against the flames as they advanced their lines, beating at beginning fires in the nearby exposure buildings with gunnysacks, brooms, and mops. If they had two lines, one hoseman would cover his head with a blanket while the other would wet him down with a stream.

Then early on Saturday, a city blacksmith found a hydrant at 20th and Church streets, which has since come to be known as the golden hydrant (it would be painted gold each year as a memorial), because it delivered the water for the final minutes of the fire in the southern section of the city. The fire had continued to burn at Lexington, at Valencia, and at Guerrero, but the firemen of Engines 19, 27, 7, 13, and 21 were coming from Van Ness and were already on 17th Street, bringing a few steam pumpers that they were pulling along the level roadway, including the Metropolitan steam engine, a huge second-class double-gated pumper. They had a strange and confident energy, and led by Chief E. F. McKittrick, Chief W. D. Waters, and Chief J. R. Maxwell, they sensed that the end of their efforts was near. They knew that the hydrant was more than nine blocks away and that their horses would not be able to make the climb, for the streets were greatly inclined going up the high hill leading to Mission Dolores Park, but there was nothing that could stop them now.

Hundreds of people had already pitched tents in the park and had come down the hill to watch the fires roll from building to building. As the firemen began to push the huge steam engine of Engine 19 up the steep ascent, they started to yell as if they were witness to the final charge in a long war. Running into the street they threw their hands on the wheels, grasped the harness poles, and tied straps to the rigging, and three hundred dislocated San Franciscans dragged the heavy apparatus up the hill, inch by inch, chocking the wheels with every progression. It took them more than an hour to get the machine to the hydrant. When the firemen finally made the hard-suction connection between the pumper and hydrant and the water charged through the line in a great burst, the crowd cheered again. The tired firemen took their exuberance to heart, and in a final surge forward ran with their hose lines toward Guerrero and 19th, and on to Valencia, and then to Lexington.

The wind began to let up, and the fires began to slow in their burning. The firemen of Engine 10 and Engine 25 found a cistern at 19th and Shotwell, a reservoir of water large enough to feed several lines of hose, four blocks east of Mission. Battalion Chief J. J. Conlon then led them in dragging the hose five blocks to fight the flames along the west side of Mission Street from 19th to 20th streets. They had barely enough hose to turn up Mission, and because they needed every inch they could find, they directed a group of policemen and volunteers to pull a twenty-foot-high street lamp right out of the ground to give them just a few more yards of clearance, which enabled them to strike at the fire on the corner of 20th. They fought at that location until they saw a spray of water shooting over the buildings from the west and Chief Conlon realized that a crew was advancing up Lexington, and that if they had enough hose, they would soon be just on the other side of 20th and Mission.

Then, suddenly, and just as unexpectedly as the first shake of the earthquake, no more flames burned before them. The fire on the south side of San Francisco was extinguished. Both Shaughnessy and Dougherty were present at the scene, and though they usually took great joy and much pride in their men when a fire was brought under control, on April 21, 1906, whatever satisfaction they might have taken in the great work of their men was trammeled by the sheer weight of the tragedy they had lived through.

CHAPTER 81

Hours before, in the afternoon, a gale wind had begun to blow, traveling across the city toward the northwest, fanning the flames of the fire created by the dynamiting of the Viavi Building. Again the North Beach section flared up and again threatened Meigg's Wharf, an area that had been judged to be under control. If the fire was to advance to the other side of Telegraph Hill it might endanger the highly flammable oil tanks and grain sheds on Pier 27, a source of combustion that could well lead to an uncontrollable situation.

To deal with the fire coming toward the bay waters, Lieutenant Freeman and the fire chiefs decided to try to prevent it from reaching Meigg's Wharf, and if that should fail, to make a last stand on East Street at the base of Pier 27.

Freeman instructed Midshipman Pond to get the *Active* under way and to follow the *Leslie* to take a position at Meigg's Wharf. Pond saw how exhausted Freeman was, and later wrote, "He looked all done in, with the sweat streaking down through the grime on his weather-beaten face onto the dirty white handkerchief he had tied around his neck, and seemed discouraged at the unfavorable turn of events. So far as I know, he had not been off his feet since we landed Wednesday morning."

As they were discussing their plan of action, Pond continued in his account, the *Perry*'s hospital steward, C. F. Ebert, reported that "he had an injured man dying in agony, and that there was no longer any morphine to ease his pain. Freeman drew his pistol and handed it to Ebert, and said, 'Here, take this and put the poor devil out of his misery.' Ebert hesitated a second, but takes the gun and walks back to the aid station."

Pond saw that Freeman was now staring at the fire with a blank expression on his face. He interrupted the skipper's thoughts and asked if he had meant what he had said. "Recovering from his dazed state, Freeman replied, 'What did I say?' Now suddenly coming to with a start, Freeman shouted, 'For God's sake, Pond, stop him!' " Pond, much relieved, dashed after Ebert and recovered Lieutenant Freeman's pistol.

As Freeman and Pond piloted the *Leslie* and the *Active* to Meigg's Wharf, Pond observed the smoke before him. "It whipped about over the city," he wrote, "for a long time, changing color alternately from black through gray to yellow as the setting sun shone on different parts of its writhing column." It was not taken as a good omen.

When they arrived at the wharf they discovered that the fire was almost upon the oil tanks and grain sheds there. They ran with their men, hoses in hand, and worked feverishly to beat back the flames. But they were too late. The fire had taken hold of the grain sheds, and as the oil tanks were ignited, they sent up massive plumes of black smoke. Freeman recognized that the danger of explosion was imminent and reluctantly ordered his men to retreat. He then lined them up at Pier 27 in military fashion, but instead of guns they held fast to their hose lines, a defense position he prayed would hold. By late dusk, the *Active* and the *Leslie* had both taken up positions on the old north waterfront, but the area seemed doomed. All the other nearby vessels, from small fishing boats to large passenger ships, began to raise their anchors, throw off their lines, and move away, as they did south of the Ferry Building, from the endangered wharves.

Through the next hours, Freeman's line at Pier 27, as well as one along Chestnut and Lombard near the now-smoldering ruins of Globe Mills, successfully held back the flames after a desperate all-night fight. But during the early hours of Saturday another problem threatened to undo all their efforts. The firefighters and sailors were greatly fatigued, and many dropped away from exhaustion. Well-meaning civilian volunteers began to take their places at the hose lines, but Freeman saw that he needed experienced men to do this work, men who were familiar with crisis, men who worked well under orders.

As if in answer to a prayer, sections of the Pacific Fleet began to arrive. Freeman consequently sent Pond in the *Active* to the USS *Chicago,* the fleet's flagship, to ask Admiral Goodrich for fresh reinforcements. Ensign Bertholf, commanding the *Active,* dropped Pond off at the *Chicago,* hurried back to the waterfront to continue firefighting efforts there. The officers on the *Chicago,* alarmed by Pond's bleak report, not to mention the soot-covered midshipman's appearance and evident exhaustion, quickly complied with his request for men, but they refused to let him return to his post until he had rested and eaten. It was now approximately 1:30 A.M. on Saturday, April 21.

While Pond was on the *Chicago,* the fight against the fire on the waterfront intensified. A sulfur works at the eastern part of the firefighting line caught fire, and the resulting fumes made breathing extremely difficult. The wind intensified and blew a greater shower of cinders, some several inches in diameter, on the men, threatening to set fresh fires. As Freeman described this scene:

> At this time my men were on the verge of collapse and were in a hysterical condition. They were too weak to handle the hose. They had been without sleep for seventy hours, and had very little food save occasional scraps they commandeered. The hardest fight we had during the entire fire was at this point. . . . The showers of cinders . . . made this place a purgatory.

But they persevered, and at last they succeeded in beating back the blaze, literally a foot at a time, until Freeman was finally convinced the fire had stabilized.

At approximately 2:00 A.M., the men from the *Chicago* arrived, several hundred strong. Ensign Bertholf and the *Active*'s crew had succeeded in checking the fire from going farther up the waterfront and were now relieved by Lieutenant Commander C. S. Morgan and a detachment of men. The *Leslie,* docked at the Belvedere ferry slip, joined by a state tug, the *Governor Irwin,* combined their strengths and threw up large volumes of water from their monitors, which then carried like a rainstorm to spray down onto the piers and the roofs of the sheds—a very practical extinguishing blanket against the burning cinders. The fire along the north waterfront was now under control, and the crew of the *Leslie* was released. On the pier, the exhausted Freeman still had a hose in his hands as a relieving naval officer approached.

So Lieutenant Freeman and the men of the *Perry,* after almost three full days on their feet, their mouths and nostrils raw with the constancy of smoke, their brows plated with the soot and grime of the heat, and their bones aching with every movement were finally and officially relieved of their duty. The great fire of 1906 on the waterfront was ultimately extinguished, seventy-four hours after the earthquake shook it into its impending fury.

CHAPTER 82

Henry Lai awoke early on Saturday morning. In his excitement he was unable to sleep, and even though exhausted from his ordeal of the previous days, he felt as good as he ever had in his life. He felt a bounty of gratitude for the old woman at the Jones Street pier who had heard him pleading, "Does anyone know Yuen Kum? Does anyone know Mrs. Cameron?"

The old woman pulled at his sleeve and whispered that Mrs. Cameron had been with the church people at Van Ness and Sacramento. Henry kissed her and ran back to that area, but by then it was night and he became lost in the unfamiliar streets. When he finally arrived at the First Presbyterian Church it was in flames and surrounded by troops and crowds. Henry panicked because he saw no Chinese anywhere, and he did not know with whom to speak. He hustled through the crowd, growing ever more desperate. That Yuen Kum had been in the church was the only information he had obtained in two days of searching, and now the church was hopelessly burning before him. Finally, he saw a man in a white collar and a black shirt, and he ran to speak with him, but the roar of the cannon and the artillery made it nearly impossible. And his English was so imperfect. Still, the minister understood and quietly told him to go to San Anselmo across the bay to Reverend Landon.

Henry had stood on line and then in the middle of a roaring, rushing crowd—thousands of people trying to leave the city. First, one boat came, then another, and Henry, try as he might, could not get close enough to be one of the lucky ones to gain passage. Finally, a sympathetic sailor opened the gate for him, not more than twelve inches, just enough for Henry to slide through. It might have been the tears in Henry's eyes that moved him to let Henry aboard or simply that look of desperation that the sailors and all of the men and women working to give order to chaos had come to see in the face of every person they helped in those days.

Donaldina Cameron was truly proud of herself and her charges that Saturday when Henry Lai and Yuen Kum finally came together. It was God's work, she must have thought, as she wrote:

> Long before the eighteenth of April the cards were out for a wedding at the Home. Yuen Kum, a clear, bright girl who had been with us several years, was to be the bride of Mr. Henry Lai of Cleveland, Ohio. The date set for the wedding was April twenty-first. And to prove the truth of the old adage "Love will find a way" let me tell you that the wedding did take place on that very date!
>
> The ceremony was performed by Dr. [Warren H.] Landon in the beautiful, ivy-covered chapel at San Anselmo, and notwithstanding all the difficulties the young man had gone through in finding his fiancée, on his arrival from the East the day of the earthquake, and all the trying experiences through which Yuen Kum had passed, they were a happy couple as they received the congratulations of those present.
>
> Just after the wedding, Mr. and Mrs. Henry Lai started for their home in Cleveland amidst showers of California roses and the best wishes of their many friends. So romance with its magic touch helped us for a time to forget our great losses.

CHAPTER 83

There was an immediate effort after the fire to reduce its economic impact on the city by controlling how information about it was disseminated. Reporters were asked to describe the disaster as a fire and not as an earthquake. Financial institutions would invest in a city that had suffered a fire, for all cities are subject to fires, but an earthquake was a different consideration altogether. Because the city needed their confidence in its building projects, writers were hired to craft positive stories, reducing the numbers in every citation of loss.

The mayor had also created a committee on history and statistics, knowing that history would demand a scrupulous account of such a critical time for the city. The banker John Drum was appointed the chair, along with two other prominent citizens. In the next few days Governor Pardee would suggest that a Berkeley professor also be placed on the committee, and as Henry Morse Stephens came aboard it was to be known simply as the Committee on History. Its information- and artifact-gathering mission was aided by the employment of two graduate students whose work was underwritten by the American Red Cross relief committee. Advertisements were placed in newspapers asking for personal accounts. Every newspaper article was catalogued, and every committee leader, ferry captain, business titan, and burned-out housewife they could find was interviewed. Time lines and maps were prepared. They worked full time for more than a year, assembling what was reported to be a great and abundant historical collection, which has since totally disappeared.

Professor Stephens, who died in 1919, would go on to be the most instrumental force in building the Bancroft Library, today one of the most efficient libraries in the world. He headed the Committee on History to record the events of World War I, and was one of the most respected academics in California. No mention is made of the San Francisco earthquake and fire papers in his own estate, and it will always remain a mystery how such an important collection came to be lost. A great fire struck Berkeley in 1923, and perhaps the collection was lost then, or perhaps it is still stored in a mislabeled box in the University of California's possession, in some library, building, warehouse, or attic of some per diem cataloguer hired long ago to build a reference file.

The fact that this collection has vanished feeds the hypothesis that it was destroyed by a group of people in whose interest it was to suppress the facts surrounding the earthquake and fire. As well as needing eastern banks to deliver sufficient construction loans, the city would also need insurance companies to cover its risks. A centralized depository of accounts and photographs that would serve as a source for an endless stream of articles and books about the collapse and burning of buildings, not to mention the deaths of so many people, did not lend itself to the confidence and optimism that the city needed to go forward into the future. It would only serve to keep the dreadful past in the present.

It has also often been suggested that the Southern Pacific Railroad, in William Herrin, sought through a cover-up to soften the impact of the crisis because of its vast landholdings and hauling business in California generally and in San Francisco particularly.

The Southern Pacific had succeeded as a business venture because the big four early on recognized the power of government in issuing land grants and building subsidies. Indeed, in 1869 the government had loaned Stanford, Hopkins, Crocker, and Huntington fifty million dollars to expand their railroad, and such was the power of the company that, in 1896, as the loan was approaching its 1899 repayment date, it petitioned for a fifty-year extension at just half a percent interest. It was the kind of presumptuous request that could only have been made by a company that took for granted its power and authority among voting legislators. The reform governor of California at the time, James Budd, was so pleased when the U.S. Congress rejected the proposal that he declared a state holiday in January 1897.

The mastermind of this and most Southern Pacific schemes was Herrin, a lawyer and expert in the newly created art of public relations. He presided over the political bureau of the Southern Pacific in the company offices at 4th and Townsend, near the sheds and yards that Lieutenant Freeman would save from destruction. Although the company had come into the hands of E. H. Harriman, its determination to control its own political destiny, and the politics of the time, continued. From 1893 to 1910, the year after Harriman died, Herrin, as the company's chief counsel, maintained control over elected officials, primarily by putting them on retainer and setting them in the direction of the company's interests. That this was patently unethical and illegal was of little concern to the company or its attorney. California historian Kevin Starr cites one reformer, a doctor named John R. Haynes, as saying: "From the village constable to the governor of the state, the final selection of the people's officials lay with Mr. Herrin."

The editor Fremont Older recalled the time he was trying to raise money to fight this kind of widespread corruption in the city. "I called on Rudolph Spreckels," he wrote in his autobiography, "and told him how matters stood. He was most enthusiastic. He . . . said, 'Older, I'll go into this! I'll put my money into this and back it to the limit. But I want one understanding—that our investigation must lead to Herrin. Herrin is the man who has corrupted our state. He is the man who has broken down the morals of thousands of our young men, who has corrupted our legislature and our courts, who has corrupted supervisors of counties, and coroners and sheriffs and judges. He is the worst influence in California. If we go into this fight, we have got to stay in it till we get him.' "

That the reformers viewed the fire as an obstacle to their efforts was indicated by Older. "San Francisco was destroyed," he wrote. "I was in the midst of a cataclysm, working as all men did in those feverish days and nights, first to save what I might of the *Bulletin,* and later to help others who needed help. But, my mind was so filled with one idea that even in the midst of fire and smoke and ruins, I thought of our plans to get Ruef and Schmitz, and mourned the delay I feared the fire had caused. I worked frantically [to deal with] this overwhelming disaster, so that we could go on with our hunt of the grafters."

Still, the Southern Pacific pursued its agenda, which was to preserve an image of a San Francisco that was safe and stable. It has been suggested that the railroad bought up every photograph of the destruction it could find to keep such disturbing images from circulation, and in the pages of its monthly magazine, *Sunset,* it ran nothing but optimistic accounts of the fire, with hardly a mention of the earthquake. Fires, of course, can be fought while there is no defense against the unpredictability of an earthquake, and the first rule of public relations is to accent the positive. (It should be noted that the insurance companies bought the rights to every photograph they could find as well, to use or suppress as evidence in denying claims.)

If this was a cover-up, it had unanticipated results because it stifled any research into earthquakes that might have been underwritten by the city, the state, or the federal government. Similarly, any pubic discussion of the hazards of living in an earthquake-prone area was discouraged. No one wanted to make too much of the quake, for no one wanted to believe it could ever happen again. It made more economic sense to claim the fire alone was responsible for the city's losses, and it did not make any economic sense to inquire how the fire had started or how it grew.

The Committee of Fifty was effective in seeing to the immediate needs of housing and feeding the more than two hundred thousand homeless, as well as reestablishing gas, electric, and transportation systems, razing dangerous buildings, and clearing the streets to make way for new construction. The railroads all participated in laying special spurs up side streets so that debris could be moved and deliveries made. Paper money was in short supply, so the committee issued clearinghouse certificates all merchants agreed to cash, in an air of focused cooperation and determination. Even as people were fleeing there was an influx of day laborers into the city seeking employment. Though they were essential in rebuilding, they brought with them a burdensome cost. Because of the hard work and overcrowding of new and untested citizens, coupled with nights of wild relief, there was an extraordinary increase in crime in the months after the fire. One laboring man, H. C. Graves, who worked in the city for the next two years, wrote about the rough-and-tumble atmosphere during the period of reconstruction: "On Howard Street near Third, an Irish lady built a frame building housing the ladies and the inevitable saloon. The ladies, 50 strong, thronged this saloon from 1 pm to 1 am. Needless to say, this place was overly popular: 5 cents for a beer, and 50 cents to $1.50 for ladies. . . ." The workers lived in places like The Phoenix rooming house, on Howard and Fourth, that "had 1600 bunks, built up in tiers of six high, in long rows," a place where there was "always a fight, an average of four an hour."

Graves also described the massive effort that went into cleaning the streets in preparation for the rebuilding. "The United Railroads in cooperation with the SP [Southern Pacific] undertook the huge task of clearing the business section of downtown San Francisco of the debris. The millions of bricks that were good were stripped of old mortar, and piled up on all the streets. The United RR built a big bunker at 1st and Mission . . . and the dump trucks drawn by horses hauled the debris up to the bunkers." Though many tons of reusable building materials were salvaged from the great mounds of the more than twenty-eight thousand collapsed, blown-up, and torn-down buildings, more than eleven million cubic yards of debris had to be pushed into the bay. This created acres and acres of additional made ground on which hundreds of new buildings were erected—once again on top of the most liquefaction-prone ground in the Bay Area, where the future collapse of buildings might in retrospect be said to be probable. But in 1906 the consideration was focused entirely on rebuilding and getting back to business.

Graves reported that all the streetlights were fixed and working by August and that hundreds of buildings were going up. Only a year later, "In the fall of 1907, you could hardly tell there was a fire of such a magnitude in San Francisco."

CHAPTER 84

Insurance companies had a mixed review of their efforts after the fire. Of the hundreds that insured property in San Francisco, only six paid what they owed, according to their policies—the Aetna Company of Hartford; California Insurance; Continental Insurance; the Liverpool, London and Globe Company; Queen Insurance of America; and Royal Insurance of England. These insurers met their obligations, according to the National Association of Credit Men, "in a spirit of liberality and honesty."

The insurance industry was not well regulated at that time, and companies were not required to keep reserves, which meant they could insure well over their ability to pay. For instance, a company like the Traders Insurance Company was allowed to write $160 million in coverage with only $1.8 million in capital. After the fire, Traders had $4,640,000 in claims and so went bankrupt, leaving its policyholders without recourse.

The Hartford Insurance Company paid off immediately, but with a 2 percent discount. Twelve other American insurers tried to meet their obligation by assessing their shareholders, but all twelve finally went bankrupt. One Austrian and three German insurers reneged on their obligations completely and withdrew from doing business in the United States. More uncertain were the many companies that honored only a percentage of their insurance promises, like the Hamburg–Bremen Company, which paid with discounts of 25 percent while at the same time being accused of much discourtesy.

The Fireman's Fund Insurance Company, which had been founded after a fire in 1863 with a provision that paid 10 percent of its profits to the widows and children of firemen who lost their lives in fires, was another of the more reputable companies. Fireman's Fund had eleven million dollars in claims and only seven million dollars in the bank, but its directors were determined to make good on their promises of insurance. To protect its policyholders Fireman's Fund found a legal remedy by declaring bankruptcy, paying fifty cents on the dollar, dissolving the company, and then giving its policyholders another fifty cents on the dollar in stock shares of a newly restructured company. It was a fair resolution, and indeed a more than generous one as the company again grew into one of the industry's leaders.

Almost all of the policyholders swore that their property damage did not occur as a result of the earthquake, realizing that they were not covered in that exigency. Most insurance policies had exclusion clauses for earthquakes, and many of the insurance companies argued that because the losses in San Francisco were due fundamentally to the earthquake, they were not liable—a position that the courts held untenable. While there was an earthquake, the protection against fire loss remained in effect. A central committee of adjusters was created, however, which determined that 10 percent of the city's damage was caused by the earthquake while 90 percent resulted from the fire and dynamiting. Consequently, the companies agreed to pay 90 percent of the face value of the claim. The committee, called the committee of five by the association of insurance companies, suggested that the city's total losses could be estimated at $1 billion, an extraordinary sum when one considers that the federal government's entire budget for the previous year was about half that sum. Other insurance experts cited a $500 million total loss—exactly the amount of federal expenditures in 1906. It was reported that in the burned districts about $235 million in insurance policies were in effect while about $180 million were paid out in claims—figures that make it obvious that almost everyone was underinsured. Still, the insurance dollars that did come into the city in the millions from Fireman's Fund and other companies were of fundamental importance to raising the city out of the ashes under the standard of its phoenix-emblazoned flag.

CHAPTER 85

In January 1906, Boss Ruef had convinced the board of supervisors, perhaps by secretly reminding them of William Tevis's and Bay Cities Water's promise of one million dollars, to stop any further consideration of Phelan's Hetch Hetchy plan, and so the legislative body of the city abandoned any claim Phelan had made for the City of San Francisco to the rights of the water of the Tuolumne River. Phelan had conveyed all of his rights in the water property to the city in 1902 and continued to argue the merits of his proposal as often as he could. But the fire had made it clear that the Spring Valley Water Company had failed them, that its failure contributed to the burning of the city, and that the city had to protect itself for the future by owning and maintaining its own water supply. While Phelan might have had some hope that his own prescient plan would be revived, Abe Ruef insisted on commissioning a comprehensive study of the water issue. When the report was published, it was not surprising that Tevis's Bay Cities Water Company was judged to be the most practical plan of the five plans that were considered.

Phelan denounced the plan publicly, and Fremont Older wrote a critical and accusatory editorial in the *Bulletin*. "Mr. Tevis," it began, "you know that by only one means—bribery—can it be brought about that the Hetch Hetchy project shall be abandoned and your scheme taken up in its lieu."

By October, so much controversy attached to the water question that the board of supervisors decided to put it on the ballot and leave it to the voters. But the looming graft investigations and trials made a plebiscite unrealistic, and it was put off.

Water was but one of the issues ripe for exploitation after the fires. Since the beginning of the year, Patrick Calhoun's United Railroads had been lobbying the influential of San Francisco to expand a trolley system powered by overhead wires, a system that Rudolph thought was both dangerous and unsightly. The Municipal Street Railways of San Francisco, the underground-wired trolley system proposed by Phelan and Rudolph, was not only more practical but would be in keeping with the City Beautiful movement. The board of supervisors had originally rejected their plan with the objection that the underground wires would short out in rainstorms.

Calhoun, who had become rich from his investments in New York real estate, oil, cotton, and manufacturing, had bought United Railroads from magnate Henry E. Huntington in 1902, capitalizing a company valued at thirty-nine million dollars with seventy-five million dollars in debt—an amount that included a twenty-million-dollar bonus for Calhoun. This was certainly an excessive amount of money and today would be entirely illegal. But Calhoun had the backing of J. P. Morgan and the Wall Street moguls in his deal, and consequently sold out the entire stock offering.

In response Rudolph and Phelan offered a proposition that the City of San Francisco could hardly turn down, for they not only promised to construct and manage the hidden-power line system and to construct drain pipes beneath the ground to prove its viability in any weather, but they also agreed to convey the line, after its worthiness was proven, back to the city at any time the city wanted, for the price of the investment plus any accrued interest. It was an opportunity for San Francisco to own its own transportation system without a penny of upfront risk.

Despite the fact that most people and the newspapers were against the overhead wires because of safety issues, Calhoun knew that in the Bay City there was an acknowledged way to gain an advantage: namely, by filling the open hands of local politicians. Through his lawyer, Tirey L. Ford, Calhoun had accordingly engaged Abe Ruef at the time he bought United Railroads to guide its business through the halls of city government. The payments by Ford to Ruef were kept secret, and Ruef described himself in later testimony as "a consulting attorney." The approval to expand the overhead wires depended solely on the eighteen members of the board of supervisors—all of whom Ruef knew intimately.

Early in 1906, Ruef's "retainer" had been increased from five hundred dollars to one thousand dollars a month, and in February, he was offered fifty thousand dollars to ensure that the United Railroads plan would be passed by the board. In protest, Ruef had replied, "The company [United Railroads] would lose respect for an attorney who would be satisfied with so little." And, finally, Ruef and Ford settled on two hundred thousand dollars as an amount that would ensure the cooperation of the board members, such amount to be paid after the approvals were made.

Even Rudolph and Phelan were offered bribes. In exchange for his support, Calhoun promised Rudolph, who owned commercial property at the corner of Sutter and Powell, to tunnel through the Powell Street hill and make that corner the busiest transfer point in the city. Phelan was offered a donation of two hundred thousand dollars by United Railroads to improve the property of Panhandle Park, one of Phelan's favorite civic improvement ventures. When both refused, the competition for approval continued into the morning of April 18, by which point everything had changed utterly.

But the earthquake, instead of opening up opportunities for making use of the Burnham plan to rebuild the city, was seized on by Abe Ruef and Mayor Schmitz as a pretext for demanding that reconstruction proceed in the quickest, most straightforward manner, a sentiment that was supported by the city's businessmen, who wanted to put the days of turmoil and tragedy behind them. Even the reforming *Bulletin* agreed with them, and in an editorial asserted, "It is desirable to have a beautiful city, but it is urgently necessary to have a city of some sort instead of a heap of ruins."

On May 14, the ordinance permitting United Railroads to create overhead wire trolleys for its entire system was published and a vote scheduled the following week. The *San Francisco Examiner* and William Randolph Hearst ran the following:

UNITED RAILROADS WOULD TRY
TO LOOT THE STRICKEN CITY

The following day they ran this statement:

SNEAK THIEVES AMONG THE RUINS AND SENTRIES WHO TURN THEIR BACKS

> ... If the supervisors are honest in this matter they can go about the solution of this problem in the right and simple way, by a short-term license. A permanent grant would be "the wrong way—a way so wrong that it smacks and smells of bribery and of a ghoulish effort to steal from the city in her time of need." Should the supervisors aid and abet this theft, the people will be warranted in setting up their effigies in lasting bronze, a group of everlasting infamy, with the inscription: THESE MEN LOOTED SAN FRANCISCO AT THE TIME OF THE GREAT FIRE OF 1906.

Nevertheless, on May 21, 1906, not a month since the fires had ended, the board of supervisors voted for the Calhoun plan of permanently expanding the system of overhead wires for United Railroads. On that same day Tirey Ford took a draft of fifty thousand dollars of Calhoun's money from the U.S. Mint, exchanged it for cash at the offices of the relief committee, and carried it in a bag to Abe Ruef. An additional fifty thousand dollars were paid in July and the final hundred thousand dollars in August. Of this sum Ruef paid eighty-five thousand dollars to the board of supervisors, and fifty thousand dollars to the mayor, reneging on his agreement to split the booty evenly with Schmitz.

CHAPTER 86

As San Francisco began to return to normal, most people attempted to go about their lives without stopping to think of their own suffering and tried to continue their normal day-to-day lives. Given that, Fremont Older's natural endeavor was to root out corruption, and only a few days after the fire, undeterred in his focus on the corrupters, he sought out Rudolph Spreckels. "Spreckels," he wrote, "had rigged up a little temporary office, roughly built of boards, in the ruins of his bank on Sansome Street. Heney and I found him there surrounded by miles of burnt brick and tangled steel girders. At once we plunged into discussion of our plans."

In planning the immediate future, the reformers knew they had to have an official mechanism to subpoena and to indict. Fortunately, the people of San Francisco had in the Union Labor Party sweep of 1905 elected an honest district attorney in William Langdon, a schoolteacher and lawyer. In October, Phelan and Older convinced Langdon to hire Heney as an assistant district attorney. Langdon had preferred a man named Hiram Johnson, a brilliant attorney who with his brother had a successful law practice and were very much progressives. Their father, however, who was one of the most important lawyers in the state, was a friend of the Southern Pacific Railroad, and so was about to become alienated from his sons. Langdon eventually hired both Heney and Johnson as assistants, and both would be the prosecuting attorneys on the many corruption trials that would take place.

Heney was determined to get to the absolute bottom of the corruption, and he told Lincoln Steffens that he would forfeit the board of supervisors if he could convict Ruef. Likewise, he would forgo Ruef if he could get Herrin. And, finally, he would do without Herrin if he could get E. H. Harriman, for he believed, as many did, that the railroad was the ultimate foundation for all the corruption in California and the city. But Heney never progressed beyond the first step in this list of targets, because Ruef was the only one of all the corrupted and corrupters whose conviction would stand the test of appeal and the only one who would serve out his time.

Heney believed that the mayor and Ruef could be implicated in a three-tiered scheme of organized corruption. First, there was police graft, which involved issuing permits for anything from hanging a sign in public to operating a boiler to serving liquor in places of questionable reputation. Franchise graft came into play with the approval of telephone systems, trolley and railroad lines, and even the applications of promoters for organized world champion boxing matches. Finally, there was rate graft, which was a constant, for setting the amount of fees and rates the gas, electric, and water companies could charge the people and businesses of the city was renewed by legislation every year or two.

In August, Schmitz went for an extended trip to Germany ostensibly to cajole the German insurance companies to meet their obligations. He expected, according to local newspapers, to be treated "as one of the crowned heads," because it was well known that he was "the man of the hour" during the emergency. Kaiser Wilhelm did not receive him, but many welcomed him with honor and ceremony. However, it was later discovered that he spent the city's money freely in the best hotels, theaters, and restaurants, and had no consideration at all from the insurance companies. While he was returning home, on November 15, 1906, Schmitz and Ruef were finally indicted on five counts of extorting money from the "French" restaurants, the Poodle Dog, Delmonico's, and Marchand's. Heney had them cold because in exchange for immunity the owners of the restaurants all testified that they had given Ruef money to maintain their licenses. On December 6, Ruef and the mayor were arraigned. So arrogant was Ruef in the confidence of his own political power that he turned his back to the judge as the charges were being read.

In February 1907, Mayor Schmitz was excused to go to Washington to meet with White House officials for a diplomatic storm had occurred with Japan when San Francisco decided to segregate the Japanese and the Chinese from its schools. The San Francisco school officials were in a quandary over the fact that some of its Japanese students were in their twenties yet were attending school with young girls in the seventh and eighth grades. Instead of ordering age limitations, however, they thought it would be beneficial to build new schools exclusively for the Asians. The ambassadors of both countries complained to President Roosevelt, and Schmitz did everything he could to reverse the order and please the president.

While Schmitz was away, Abe Ruef pled not guilty to all the charges brought against him, and it was reported that Judge J.C.B. Hebbard was so drunk at the hearing that it was uncertain if it had been conducted in the correct legal manner. Ruef was therefore ordered to appear back in court in two days but failed to show up. The police chief and his men went searching for the boss, but after two days failed to locate him.

William Burns was then asked to look into it, and he and the prison warden, William Biggy, found Ruef in little over an hour, hiding out in a roadhouse just a few miles from the city. Ruef was beginning to realize that he was entering difficult waters, and his brief disappearance had given him time to formulate a plan.

His dilemma was to determine if he should ally himself with the mayor or distance himself from a man in whose intellect he had no confidence. In an essay Ruef later wrote he claimed that he wanted, but was unable, in the previous election to separate himself from Eugene Schmitz and from the labor union rabble around him, which "would eat the paint off a house." This statement would be largely interpreted as describing the board of supervisors, about whom it was more demonstrably true, but Ruef had, in fact, directed it against the Labor Unionists in the mayor's circle, the very people who had granted him such powerful status.

In March 1907, Burns had exacted, with offers of complete immunity, confessions from each of the board of supervisors. The seeds of his and Heney's hard work began to bloom, and for paying graft to the board of supervisors on February 23 and 24, 1906, Louis Glass, the general manager of the Pacific States Telephone Company, and Theodore V. Halsey, its political representative, were indicted. The two men would be found guilty in the summer of 1907, but the process of appeal was a long one, delayed several times by the emergency medical conditions of the defendants. Finally their conviction would be reversed on an esoteric technical matter in November 1910, and all other charges against them dropped in 1912. They would continue as first and second vice presidents of the company for many years.

On May 7, 1907, Ruef made a confession to Burns that Tirey L. Ford had given him a total of two hundred thousand dollars for the approval of the expansion of the overhead wire system for the trolleys. Burns was overjoyed, but when he and Heney examined the statement more closely they concluded that the political boss did not tell them anything they did not already know. What Ruef had omitted to acknowledge was the understanding between him and Ford that some of the money would go in payoffs to the board of supervisors. He would later again claim that the money had been intended as legal fees to retain his services as a lawyer.

In May, the grand jury returned fourteen indictments against four citizens for bribing the board of supervisors in connection with the trolley franchise: Ruef, Schmitz, Calhoun, and Mullally. Ford was indicted separately for delivering the money to Ruef. Fourteen indictments were brought as well against Ruef, Schmitz, and three gas company executives—Frank Drum, Eugene de Sabla, and John Martin–for the bribing of the supervisors in setting the gas rates. Also, A. H. Umbsen, J. E. Green, W. I. Brobeck, and Ruef were indicted for the thirty thousand dollars that had been passed to Ruef to get a rail line approved to a Parkside real estate development on 19th Street near the Presidio. Ruef's defense in each case was that these payments were perfectly legal retainer lees for his legal advice, and Schmitz simply denied receiving the money.

Patrick Calhoun ultimately took the stand and pleaded the Fifth Amendment, as did Ford, both men arguing that it was not their responsibility to defend themselves but rather the responsibility of the *prosecutors* to try to find wrongdoing. Heney, of course, had the confessions of the board of supervisors, but without the corroboration of Ruef there was no way to connect the money that Ruef had given them either to Ford or to Calhoun.

In that same month, however, on May 8, 1907, Ruef made a deal with Heney and was given limited immunity from all indictments but those connected to the "French" restaurants to tell the full story. The immunity contract was negotiated by Ruef's good friend and counselor Rabbi Bernard Kaplan. There had been much talk of anti-Semitism in the Ruef allegations, and it can be said that as a group the Jews of San Francisco were in support of the defense. They were proud of Abe Ruef, a prodigy who spoke many languages and who rose in political influence in the Labor Union Party and to prominence in their fair city. The controversy did reach as far as New York, however, where the famous Rabbi Stephen Wise was to write, "Israel is not responsible for Ruef's crimes. . . . Israel is unutterably pained by this blot upon its record of good citizenship in America."

On June 13, 1907, Mayor Schmitz was convicted of extorting money from the "French" restaurants, and consequently the board of supervisors elected an interim mayor from its own ranks—an election based not on the fact that the convicted mayor was in jail but simply that his absence rendered him "unable to perform his duties." The confessed bribe taker James Gallagher accordingly took office and remained there for a month until the sentencing of Schmitz. Once Schmitz was sentenced, however, the city realized it needed a more permanent officeholder in city hall, and the board of supervisors elected fellow supervisor Charles Boxton to the position, until the next formal election. In the following few days, as almost all of the board of supervisors resigned, Boxton was kept busy learning the mechanics of his new job, while at the same time testifying with immunity at the trial of telephone executives Glass and Halsey about taking their bribes as a supervisor. As the *San Francisco Examiner* observed acerbically, "Having put our bribe-taking Mayor in jail, and having put in his place a taker of smaller bribes, we have now substituted for Gallagher, [this man] Boxton, who differs . . . principally in having sold his vote for still less of the bribing corporations' money."

As a mayor, the naïve Boxton was something of a political catastrophe. He took offices far from city hall, and then fired any secretary or member of the support staff who did not show up at the new address. Phelan and Rudolph persuaded the board of supervisors to consider another candidate, and Boxton resigned after just one week. In a gesture toward honesty the supervisors then elected Dr. Edward Robeson Taylor, an eminent physician and lawyer.

As in most great comedies an underlying sadness pervaded the playing-out of these scenarios. For Rudolph and Phelan, finally, there did not seem to be a way out of the corruption in San Francisco, so mired was the city in it. They had taken the public position that the corrupt politicians were just 50 percent of the problem and that fairness dictated that those guilty among the city's leading businessmen be brought to justice as well. But needless to say, San Francisco's social and business leaders perceived the issue differently. For them, it was the cost of doing business in the city and was a victimless crime. As Calhoun, Glass, Halsey, Ford, Mullally, and others who were members of their own clubs faced prosecution, other businessmen in the city began to ask themselves if they would be next. People began to believe that the corruption investigations were not in the best long-term interests of the city. Hardly any businessman did not have something questionable in his past, whether it was a ticket, a gift, or a trip that had been paid for to foster a friendship with a politician, and so Rudolph and Phelan began to feel the consequences of an ever-increasing ostracism. The support for the moral drive that had given the prosecutions appeal suddenly and inexplicably evaporated. The city was prospering, and its people just wanted to get on with it.

Men who were thought to be friends would get up from their tables and move away as James Phelan approached them in the Bohemian, and Rudolph, besides having been requested to resign from the Pacific Union Club, was asked for the good of the First National Bank to leave the office of its presidency. While he would not permit himself to be forced out of the Pacific Union, he did step down from the bank's board, though still retaining his stock. Fremont Older simply resigned from the Bohemian, but his and Rudolph's wives complained that they had been dropped from everyone's list and most nights were consequently spent at home.

Still, the reformers could see that change was in the air. Despite the closing of ranks of San Francisco's business community, it had become obvious that people would no longer think that corruption was simply another business expense. And to Phelan's and Rudolph's delight, the larger issue of the Southern Pacific's pervasive and corrupting influence in California's politics was also being met. In 1907, the Lincoln–Roosevelt League was started to create opposition to every influence on the State of California's business by the Southern Pacific. Rudolph was the president of the San Francisco chapter. The organization developed such a strong membership that, in 1910, Hiram Johnson, fresh from the corruption trials, would be able to win the governorship. His single campaign promise was "to kick the Southern Pacific out of politics."

CHAPTER 87

While Heney, Johnson, and Langdon were successful in the trials, the appeals took years to go through the courts, until, finally, all but Ruef's conviction were overturned.

When Ruef had promised to confess all for a limited immunity in May 1907, Heney had been confident that his testimony would provide the missing link that connected the graft receivers with the graft-giving executives. But a week later, Ruef fired his two attorneys and, some say on the advice of Rabbi Kaplan, reneged on his contract for immunity, declaring once again that he simply received attorney's fees for legal advice. On the one charge that Heney had kept out of the immunity agreement, the charge of extortion from the "French restaurants," Ruef surprisingly changed his plea from not guilty to guilty.

Heney was furious and canceled the immunity immediately, deciding then and there to prosecute Ruef for bribing one of the board of supervisors. Of the eighteen who had admitted taking bribes he chose Supervisor John J. Furey.

It is probable that Ruef's legal situation was much worsened by his backing out of his immunity contract since lawyers, administrators, and judges have long memories and are naturally biased against insincerity of any type—particularly when parole applications are made. And, ironically, the charge to which Ruef pled guilty was the same charge to which Schmitz had pled not guilty—a charge that would later be cited in Schmitz's conviction reversal not to be a crime at all.

When the mayor was convicted of extortion in June he was remanded to the city prison to await sentencing. His conviction automatically stripped him of his office, but his friendships endured. William Biggy who was then called an *elisor* (jailer), made room for Schmitz in his own home for the several weeks between conviction and sentencing. But on July 8, 1907, Schmitz was sentenced to five years in the state penitentiary at San Quentin. The presiding judge, Frank Dunne, launched a long attack on his character, warning, "You will lose the respect and esteem of all good men." He was met with constant interruption from Schmitz, who predicted that the decision would be reversed and that the people would reelect him as their mayor.

Schmitz waited out the time of his appeal in the county jail and never did have to walk the halls of San Quentin. After nearly seven months of incarceration, the district court of appeals of San Francisco reversed his conviction, and Eugene Schmitz was freed on January 9, 1908. The finding ruled that *extortion* implies the demanding of money against the threat of harm, but that anyone could apply to the police chief—for themselves or for a house of prostitution, for that matter—and ask that a license be given or denied. Not only was there no threat involved, decreed the court, but it also stated that the charges failed to list the mayor's occupation as mayor, a fact that would have given him an ethical but not a legal reason to keep from representing the interests of a "French" restaurant.

Schmitz thereafter entered into several unsuccessful business deals, and then ran again for mayor in 1915 as a down-and-out underdog against a popular mayor, James Rolph. Schmitz always retained his charm and received thirty-five thousand votes in his losing bid. However, in 1917 he ran for supervisor and was elected to a series of two-year terms, always maintaining that he was made an unwitting martyr by the fabrications of Abe Ruef.

Shortly after Ruef's trial finally began in August 1908, after seventy-two days of jury selection, a man by the name of Morris Haas walked into the courtroom on November 13 and shot Francis J. Heney at close range. Seven months earlier, Haas had been a potential juror, but Heney discovered that he had been arrested for embezzlement twenty years before. In fact, twenty years previously the governor had pardoned Haas, who had gone on to head a large family, succeed in business, and in every respect be an exemplary citizen. But he forfeited that status forever when the bullet went through Heney's mouth, ear, and head. The shot was not fatal, however, and after several weeks in the hospital Heney was miraculously strong enough to visit the courtroom. In yet another event that has never been fully investigated, several weeks later Haas was found in his prison cell with a gun at his side, shot dead either by his own hand or by another's.

In December 1908, Ruef was finally convicted of bribing Supervisor Furey in exchange for his vote for the overhead trolley wire application and was sentenced to fourteen years. The net time for this sentence was nine years, dependent on the consistent good behavior of the prisoner, and the law allowed a parole consideration after serving half of this period. In the end, Ruef served a year in the county jail waiting for his several appeals to find their courses to denial before he was delivered to San Quentin, where he was incarcerated from March 7, 1911 to August 23, 1915, serving a total confinement of nearly five and a half years including his year in the county jail.

During this time editor Fremont Older did a remarkable turnabout. As he wrote, "I have tried to repent for the bitterness of spirit, the ignorance I displayed in pursuing the man Ruef, instead of attacking the wrong standards of society and a system which makes Ruefs inevitable. . . ." He went on to publish Ruef's memoirs and wrote appeals for letter-writing campaigns to have the former political boss released.

A few months after Ruef was paroled, the disbarred lawyer opened an office on Montgomery Street. The sign on his door read: A. RUEF, IDEAS, INVESTMENTS, AND REAL ESTATE. One of his ideas was to create a café on the property he owned at Fisherman's Wharf and surround it with colorful stores and bizarre seafood shops and restaurants—surely a million-dollar idea. He did have more than a million dollars when he was arrested, and after his legal fees were paid still had a half million when he was finally released from prison. However, his real estate holdings did not do very well, and when he died in 1938, after the down years of the Depression, he was bankrupt.

CHAPTER 88

Patrick Calhoun had seen how Henry Huntington had made millions by buying large tracts of land around Los Angeles and then increased the value of the land by building railroads to reach it. Calhoun made similar investments in Solano County, just east of Napa, and was determined to build a railroad to get there. But in raising the money for land acquisition, his investments in America's big city railroad systems became over-leveraged, and the New York bankers who held large positions of stock in his company forced him out of the presidency of United Railroads. With no railroad to control, his land investments declined in value, and by 1916 Calhoun had gone bankrupt.

In a 1931 book, Lincoln Steffens erroneously declared that Calhoun had died, and the still-whinnying Calhoun threatened to sue the publishing company if it did not recall its books, which it ultimately agreed to do. Even in his old age, Calhoun saw opportunities in California, and he returned at eighty years of age to invest in newly discovered oil fields, where he revived his dwindling assets to create new and substantial wealth for himself. But seven years later, still in good health and returning from a party where it was supposed he had been drinking liberally, he was hit by a car while walking across a street and died instantly.

CHAPTER 89

Frederick Funston died in San Antonio, Texas, in 1917, when his heart gave out at the age of fifty-one. The general's persona had been much elevated in the eyes of his countrymen, so much so that at his death adulating Texans opened up the gates of the Alamo for the first time in its history so that the body of a great American could lie in state for the public to view. In any reading of his life, there is little doubt that he was a great American, a bold adventurer, and a man of courage. For those four fateful days in San Francisco, however, he entered into a class of leaders who must be judged as good men who made bad decisions. Throughout history many people have shared that fate, for mistakes can be made in the most perfunctory, deliberate, and confident way when men and women undertake to lead in areas of uncertainty and newness, when they have unreliable information or lack the necessary education and experience. In consequence, people can perish and cities can burn.

Today in San Francisco, there are many honorifics for General Funston, with Fort Funston and Funston Avenue, and in city hall, there is a large and imposing bust of him.

However, there is no honor anywhere to be found for Lieutenant Freeman and no commemorating statue of the firemen of 1906.

CHAPTER 90

Claus Spreckels suffered extraordinary real estate losses in the fire, including his mansion, his newspaper, the Spreckels Building, and many other buildings he owned in the burned-out areas. Even his original business, the Spreckels Market Fruit Company at 727 Market Street, which he had long ago sold, was destroyed. Inspired, perhaps, by the pioneering San Francisco spirit, he was said to have exclaimed, "I can be a chauffeur." His gallant wife then added, "And I can sell my embroidery."

His fortune, actually, was estimated to have declined from more than thirty-five million dollars to less than ten million dollars, including the thirteen million dollars each he gave his older sons John D. and Adolph. Claus would die less than two years later, at the age of eighty, on December 26, 1908, once again on good terms with all of his children. Rudolph and his brother John D. were at their father's bedside when pneumonia stole his last breath and as his other children and grandchildren rushed to the family home. Obituaries would all mention the fact that he had broken the sugar trust of the East Coast by building in Philadelphia the largest sugar refinery in the world.

His wife, Anna, must have suffered through the many years of embarrassing press stories about her family, but she had a mother's courage and optimism, and when she died in 1910, her own obituary stated, ". . . during the estrangement between father and children their mother was anxious to bring them together so that the cast-off sons and daughter might participate in the joint wealth of her husband and self, share and share alike. It is intimated that the reconciliation between Gus, Rudolph, and Emma with their father was a direct result of the efforts of their mother."

Rudolph, like his friend James D. Phelan, who went on to be a U.S. senator in 1915, maintained his early successes. He was offered but declined the post of ambassador to Germany by President Wilson in 1913, and he squelched a movement to make him a U.S. senator, keeping the promise he made during the graft period to never seek or run for public office. By 1928, Rudolph was one of the wealthiest men in America, with an estimated worth of fifty million dollars (about a billion dollars in today's currency). In 1928 alone he was said to have made eighteen million dollars, mostly from an investment in the Kolster Radio Company.

He moved to New York City, separating for a period from Nell and living in the Plaza Hotel, and became a man of great consequence. He belonged to every "right" club: Union Pacific, Bohemian, University, Burlingame, Metropolitan (in Washington), Downtown, Bankers, Turf and Field, and Embassy. But during the stock market crash of October and November 1929, Rudolph did all he could by buying and supporting stocks of his own companies and those of his friends, and he lost substantially. In 1931, he was sued by the radio corporation for stock manipulation, but the case was thrown out of court. In 1932, he was forced to put his principal company, the Spreckels Sugar Company of Yonkers, into receivership, claiming that it was a good-standing and well-managed company but lacked liquid capital to pay its bills. The following year, in trying to recoup his losses, he invested heavily with disappointing results, and thus began the slow but certain slide down the slope of ruin, culminating in a bill from the Internal Revenue Service for $1,402,042, due to his windfall profit in early 1928.

Rudolph attempted to restructure his company in 1934, but it finally went into bankruptcy as tax liens were placed against it. The loss of the Spreckels Sugar Company was an extraordinary setback, one from which he never recovered. His had become another American story of making millions and losing millions. It is said in the family, though, that while walking one spring day along Fifth Avenue with his daughter Anna Claudine, the child who had survived the fire to be born in Oakland and who was named for Rudolph's mother and father, Rudolph explained that in all things there are just two choices. The question is not whether it should be left or right, up or down, invest or not invest, marry or not marry, stay or go. It is always, simply, a matter of what is right and what is wrong. "Some financiers," he is quoted as saying, "jumped out of windows when they found themselves bankrupt. I never lost a night's sleep over it." He was in the long run more a man of principle than a man of money.

Rudolph Spreckels lived until 1958, surviving Nell by nine years. His wife left him a small legacy, which enabled him to live in a small, ground-floor apartment in Burlingame, to eat his favorite sardines on toast for lunch, to drive an old Chevrolet, and to continue an unobtrusive social life far from the excitement of his past. To the end he was a man who believed that his was a life well lived.

The disappointments of the corruption trials were great, but Rudolph never wavered in his belief that he had done absolutely the right things in supporting and financing the investigations that turned citizen against citizen in the upper reaches of San Francisco's society. And he had never forgotten that the entire faculty of the University of California, with the approval of its president, had sent him a letter expressing "appreciation of the great work already accomplished [in the trials]," and hope that his reforming work would be "carried on to the end."

CHAPTER 91

On Saturday, after a short rest in the back of a department hose wagon, Jack Murray determined to see to his family, department regulations or no regulations. He decided first to tell Captain Tom Murphy that he was going to take a leave, for it was a matter of respect. They had been through so much in these last few days that Jack did not want Murphy to be overly concerned about him. It would be three years before Tom Murphy would become the chief of department, but he would never lose his sense of fairness. He told Jack that another fireman had already gone absent and that he was powerless to stop him. He said something to the effect that he did not know anything except that when it was time to write the names of the firemen on duty in the roll call book he could write only the names of the firemen who were present. The department had given no formal order to relieve the firemen to see to their families, even those like Jack, whose firehouse was one of the twenty-one that burned to the ground, putting twenty-nine fire companies out of work. Jack had lost everything that connected him to his work and family—his dress uniforms, his extra work vests, his duty and dress caps, photographs of Annie and the children that could never be replaced, and all those books on hydrolics, department regulations, and management guidelines—the study material that would help him pass the test for the rank of lieutenant.

The department would eventually be forced to lay off many of the firemen who no longer had a firehouse to go to, and it would rebuild itself slowly as the city itself rebuilt. But Jack was one of the lucky ones, because he had just enough time in grade to be transferred to Engine 21 at 1152 Oak Street. But on that Saturday, he decided to take the risk of departmental charges and go to Annie. He knew that his family was staying with relatives in Oakland, and he cleaned up as best he could and took the ferry across. There was still no charge for the transportation, and even in his weariness Jack felt a delight in traveling for free, and also because he would soon see his wife, his girls, and young Will. He knew the stories he had for young Will would make him feel much better about the loss of his fire truck.

But when they were finally reunited he discovered that his relatives had treated the family as if they were ordinary refugees. They had been relegated to a tent in the backyard, and the children had had to sleep on blankets that had been spread across the bare ground. Jack was not a man who was quick to anger, but he wanted so much more for his family, and he expected so much more from their own relatives. His children should have been given the beds of the hosts if they knew anything at all about Irish hospitality. He told Annie to pack up immediately, that they would go to Vallejo, to his mother's home, where they would be treated more like a family that had been knocked around but not knocked down by the fire, a family that was kept together by the kind of love and respect they expected from their own flesh and blood.

While living with her in-laws Annie kept her family busy and in clean and ironed clothes until Jack found them a modest attached house in the Western Addition. Of all the objects that had been left behind in the trunk of mementos Annie missed her silverware most. It would be expensive, maybe too expensive to ever replace it. Only a set of two serving forks survived, mispacked in one of the duffels. That represented the Murray fortune: two silver serving forks, the bedding, and the clothes they carried from Telegraph Hill.

Jack Murray was later brought up on departmental charges, as were several hundred other firemen, for going absent without approval. It was a cold-hearted decision, but Jack was more than willing to take any fine they might mete out. There was no price on the joy he had felt that day in reclaiming his family.

But Chief Shaughnessy, the new chief of department, rescinded all charges, and Jack went on to become a lieutenant, with a much respected thirty-year career. He was the proudest man in history when his son Will was given the badge of the SFFD, and then again when Will was promoted to lieutenant. It was a beautiful sight to see them both together wearing their uniforms, for the father and son represented the brotherhood of the fire department like no others. They were friends first, and about that there was no doubt. It was a great regret of Will Murray's and indeed of all their friends' that his father did not live to see him become the chief of the San Francisco Fire Department, a position he held from 1956 to 1971.

They were together at the firemen's ball in 1938, a great annual activity with entertainment of song and feats provided by the firefighters themselves. Will was the master of ceremonies and kept the crowd in laughter between events. Also present was Will's young son, Bill, who would become the third Murray in the SFFD, reaching the rank of chief, and then serving as volunteer chief of the Glen Ellen Fire District in Sonoma County.

Suddenly, Jack Murray fell to the floor, in the presence of his family and all the firefighters he loved so much. The men of the newly established rescue company worked on him for an hour before a doctor finally told them to rest. Jack Murray's heart had simply given out, and as he lay on the floor, he opened his eyes just once, and he saw the faces of his son and grandson. It was the best last view any man could hope to have.

CHAPTER 92

The officers from the *Perry,* Bertholf, Pond, and Freeman, were naval men to the core. Ensign Wallace Bertholf would eventually rise in rank to command the naval transport *Harrisburg,* in World War I. He would be honored for bravery, earning the coveted Navy Cross in a sea fight during the war. He retired from the navy in the rank of captain, but he left the service a legacy in his son, born in San Francisco, whom he conspicuously named Mariner. Mariner would also reach the rank of captain in the U.S. Navy, as would *his* son, Charles Mariner Bertholf. All four Bertholfs now rest side by side at Arlington National Cemetery.

In 1906, when Pond was a midshipman, he was certain that he would continue in the navy like his admiral father, and he considered himself fortunate that after many years of service he finally did, like Freeman, reach the rank of commander. In 1921, however, he was forced to retire prematurely due to the onset of deafness that had begun to afflict him when he served as a gunnery officer. He would live to see two of his three sons follow his family's naval tradition, attend Annapolis, and eventually reach the rank of captain, and to see his third son become a colonel in the air force.

John Pond lived in Berkeley in his retirement years while his old skipper lived just south of the city in San Mateo. Nellie Freeman had died at the young age of fifty-one, and since there were no children, Frisky was left alone. Fortunately, when on active duty, he had bought a house, which provided at least something of a security, for he had no pension. To make ends meet, Freeman decided that he would rent out his home and use the slim income from that to sustain himself in an inexpensive hotel room in the small Salinas Valley town of Soledad.

Pond and Freeman had not kept in touch, and they met quite by accident one day in 1939, almost thirty-five years from their great effort as firefighters. Pond was shocked to see the haggard and frail appearance of his always-robust former captain and stunned to disbelief by the story he was told. It was unthinkable that his friend and superior officer would have gone on to live a life of disgrace, unhappiness, and dishonor.

How could that have happened? He remembered what he himself had written about Freeman in his notes on the fire: ". . . he [Freeman] was not the type of man who would wait for instructions before taking action in an emergency. He was a born leader of men, a skipper whose men would go *to hell and back* [emphasis added] for him."

Pond could only conclude that to bring Captain Freeman to such a lowly end someone must have been determined to undermine his interests along the way. He then resolved to find out more about the particulars of Freeman's discharge.

As a retired naval captain and the son of an admiral, Pond had much respect in places like the naval academy and the naval archives office. He soon discovered that Freeman's offenses were routine in nature and at most should have led to a reprimand, perhaps a demotion and an opportunity to retire. It was hard for Pond to reconcile these accusations with the authority of Freeman's voice as they pushed forward into the many fire-filled streets and alleyways of San Francisco.

Pond remembered that not long after the fire, he had seen a letter on the *Preble*'s bridge desk, a letter Freeman never mentioned but allowed him to read. It was from Senator Albert Beveridge of Indiana to the secretary of the navy, asking that the navy promote Lieutenant Freeman for his work during the crisis, described in the restrained tones of the day as "eminently satisfactory." It was a significant compliment for Freeman, and in Pond's view, deserved. Pond had discovered similar encomia while examining the record. On April 21, 1913, when Freeman was assigned as a lieutenant commander to the USS *Connecticut,* he was awarded another commendation, one of many. A sailor had fallen overboard into the cold waters of the sea at Philadelphia, and without thought to his own safety, Freeman immediately slid down a fall rope to the water and brought the end of the rope to the flailing man, calming him until they were both pulled to safety. He enjoyed more than positive annual reviews and evaluations by superior officers, which were consistent in the ten years following the fire and was cited several times for heroic action. Freeman had been singled out early in his career in his time at the naval academy at Annapolis, class of 1895. In his second year there, studying engineering, he was awarded a lifesaving medal. Though there are no particulars, it is reported he was given a "commendation for saving a life."

Freeman's downfall was caused by what would later be diagnosed as a deep depression, manifested by an addiction to alcohol. In the navy of World War I, though, drunkenness was a serious offense, a weakness of will, and without explanation several officers inserted notes into his personnel record describing Freeman's drinking. There was never an accusation of wrongdoing so much as a criticism of his demeanor and observations of the noticeable "odor" of alcohol.

During the war, commanding the destroyer USS *Corsair,* he escorted across the Atlantic a convoy of seven yachts that had been conscripted by the government to be used as merchant ships. Though they encountered terrible gales, the flotilla made it safely through to Brest, France, where they went on patrol and transport duty.

In France, Freeman was made commander of a squadron of the four biggest yachts. These vessels were all armed sufficiently to destroy submarines, and on October 17, 1917, while escorting the military transport steamer SS *Antilles,* one of Freeman's flotilla, the *Aphrodite,* developed engine trouble and could not keep up with the transport. Freeman released her from the convoy, and she was separated from the group. Not long afterward, a German submarine, U62, hit its mark on the 6,878-ton *Antilles,* and it went down. Freeman led the rescue efforts and was commended for saving 118 men in the frigid waters of the Atlantic, though 67 were tragically killed.

Still, some might have speculated that had the *Aphrodite* not been sent back by Freeman it might have saved the *Antilles*. Freeman went on to say of his actions, "This happens when any military commander acting independently must make the decision. It was reported by radio to [Admiral] Fletcher. The *Aphrodite* would have been separated anyway."

This sinking caused great consternation in the navy. Freeman was praised by Rear Admiral W. B. Fletcher, the commander of the U.S. patrol squadrons in France, for an "admirable . . . work of rescue." However, the admiral also noted in his report on Freeman's fitness, that the officer had been notified previous to the *Antilles* sinking that "he had been indulging too much in intoxicating liquors," and that in regard to the *Antilles* he had been remiss in letting the *Aphrodite* become separated from the convoy. But the commanders of the convoy vessels assured Freeman that "this convoy was the best managed and handled of any they had been in up until that time," and the court of inquiry investigating the loss of the *Antilles* had "no fault to find" with Freeman.

It should be noted here that just ten weeks earlier, none other than Franklin Delano Roosevelt, then assistant secretary of the navy, wrote in a spontaneous and unrequested evaluation, "Commander Freeman has during this period under my office, and as senior member of the special Board for Patrol Vessels has shown excellent initiative and an ability to accomplish results promptly, and performed his duties to my satisfaction."

The navy reacted harshly to the unfortunate incident, however, and held Admiral Fletcher accountable for the loss of the transport steamer and in humiliating circumstances transferred him to lesser duty. On November 2, 1917, Rear Admiral W. B. Wilson arrived in Brest to assume the command. He could not have been inclined to be sympathetic to Freeman, a man whom he must have seen as causing the relief of an admiral. Unfortunately, Freeman almost immediately became involved in a confusion about scheduling and did not report for duty when he was expected. He was relieved of duty as commander of the fleet of yachts, and in his formal report on the incident Admiral Wilson described Freeman as having "the appearance of one who was recovering from a protracted indulgence of intoxicants. His face was flush, and to a certain extent, bloated. He was very nervous, and his mode of speech very different from his usual one. [Wilson had met Freeman only once previously for a few minutes.] I said to him that it was my opinion that he had been on a debauch . . . he made no denial or response."

Freeman's only response was, "I hope the Admiral will give me another chance." Wilson, however, absolutely refused, and informed Freeman that "he would, in the course of the next few hours, receive orders to go home." And so, in the course of forty-eight hours of Admiral Wilson's arrival, Commander Freeman had been accused, tried, found guilty, and sentenced to relief of command and transfer, all informally executed without any adherence to code, the Uniform Code of Military Justice, or moral authority.

The following day, November 5, the convoy left the port of Brest without him. At about 1:45 A.M. that night the ships had traveled out about seventy-five miles from the coastline of France, and there they were set upon by the German submarine UC71. The *Alcedo,* at 983 tons and 275 feet long, a yacht that the navy had bought from the famous socialite banker George Drexel, went down in eight minutes. Many sailors were saved, but twenty seamen and one officer were drowned.

This was devastating news for Freeman because he had arranged for the acquisition of that vessel, and had had it transported to New York's 57th Street Pier to be made ready for battle. And he knew many of those young sailors who had given their lives.

In gathering whatever facts about the case his friend John Pond could put together, it did not take him long to realize that the loss of twenty-one sailors on the *Alcedo* must have weighed heavily on the psyche of his old skipper. Aside from knowing most of them personally, he had also read the official report from Admiral Wilson, which said that "the *Alcedo* failed to maintain her assigned position as flank guard of the convoy. There was a lack of alertness on the part of her lookouts. She violated the Squadron's doctrine in failing to attack [in the time between sighting the submarine and being hit by the torpedo]." Freeman must have thought, as any man who had given his life to the navy would have thought, that if he had been there in command he might have been able to make a difference.

In November 1917, Freeman was attached to the Fourth Naval District in Philadelphia. It was soon after his arrival that he fell into a deep depression, and what had obviously been a drinking problem worsened. It seems that Freeman began to wallow in alcohol.

By the following year he was brought up for court-martial, with charges that included one absence from duty, being out of uniform three times, and drunkenness four times. The absence was the most serious accusation, but it is clear that soon after arriving in Philadelphia he must have suffered a complete mental and physical collapse.

Commander Freeman was court-martialed, found guilty, and ordered dismissed in an order of May 27, 1918. It was not his fate to be shown lenity, not to mention mercy, and the third charge of being out of uniform of which he was acquitted was immediately overturned by the review. He was dishonorably discharged and so deprived of all privileges of a citizen of the United States. He could not vote. He could not participate in government programs or veterans' benefits. He was on his own to live out his disgrace.

Undoubtedly, there were times that Frederick Freeman just wanted to go back to San Francisco and to live there in peace. From all accounts he was an introspective man, and he knew that he was a good man with a drinking problem. He had requested help in the navy's hospitals and was denied. On April 2, 1918, he had requested retirement and was denied. He was not given a special court-martial, usual for officers charged with misdemeanors, but a general court-martial, which was reserved for seamen and officers charged with felonies like murder, rape, and theft. He had committed no crime or indecency against another person, had harmed no one in his career of more than twenty years in service to his country. He had risked his life countless times to help others and was commended often. People who worked with him loved him. But a small part of his heart was consumed with the kind of devils that for centuries have led men and women to alcohol. He had been overshadowed by a general who did not want any action other than his own recognized in the rescue efforts of the great earthquake and fire but, most significant, he was tormented by the loss of so many who went down into the sea as the *Antilles* sank, and those twenty-one young friends who went down with the *Alcedo*.

This is the stuff of American history. A hero, like the refinement of character, is molded from within and not from without, not created by the force of a military order or a newspaper article. But sometimes men can be erroneously marked, misidentified, and misfiled, and so it was with Frederick Freeman.

CHAPTER 93

His old friend began to stir in his bed, and John Pond was startled from his reverie. He had been thinking of those last twenty-two years that Commander Freeman had lived quietly and without complaint, accepting his punishment with the strength that was at the core of his heart. The midday sun was now illuminating the room and brightening the commander's face. He had never sought out old friends to ask their help, or gone to a military appeals attorney. He held his burdens inside himself, stoically, though they were drawn across his face. It was undoubtedly a tired face, a temperate but worn face, and it was also one that wore the dragging lassitude and unresponsiveness that comes with fighting pain and the relentlessness of being sick.

"John," the skipper said, "so good of you to come."

Pond rose and went to the side of the bed. He put his hand inside his coat and brought out the letter. "I have news," he said. "Are you feeling okay?"

"Fair in fair weather."

"Good," Pond replied, opening the folded page. "This letter could not be sent to you, but it was sent to the governor of the State of California, and they sent me a copy for you. May I read it?"

"Please."

"It begins," Pond read, "January 25, 1941. My dear Governor Olson. It is my pleasure to inform you that on January 23, 1941, I granted a full and unconditional pardon to former Commander Frederick N. Freeman, U.S. Navy, after considering the information and recommendation contained in your letter of January 16, 1941 . . ."

Pond looked up, and in the sunlight he saw the moisture glistening in Frisky Freeman's eyes. "It's from the White House, Commander," Pond said. "It's signed by Franklin Delano Roosevelt."

For several years, John Pond had worked to overturn what he saw as an unfair court-martial verdict of his former skipper. He gathered the facts in Washington. He wrote letters to California's governor, the city's mayor, old shipmates, and many others, pleading for their assistance in obtaining a presidential pardon for Freeman. He gathered testimonials, from the U.S. Naval Academy Graduates Association, from retired superior officers and enlisted men. Not long into his work, Pond's mission took on a sudden and special urgency, for he learned from Freeman's doctor that his old commander was dying of cancer. He had only weeks.

The intercession of California's governor, Culbert Olson, and San Francisco's mayor, Angelo Rossi, combined with Pond's stirring account of Freeman's heroic work apparently found a sympathetic ear in President Franklin D. Roosevelt, a former assistant secretary of the navy. The president's naval aide wrote, "The President was glad to assist in this particular case . . . and was very pleased to learn that his action brought happiness to 'Frisky' and others." Because of this familiarity in referring to Freeman as "Frisky," Roosevelt must have remembered Freeman from the promotion endorsement he had written for him while assistant secretary of the navy, and perhaps even from his signature on the subsequent court-martial endorsements. He quickly granted the pardon, which was issued just a few weeks before Freeman's death.

No words were spoken as the two men faced each other. They both could feel, certainly, a curtain descending on a saga that was not a happy one. It was a story that had taken a tragic turn more than two decades before, and now, finally, it was given, by the act of a president of the United States, a gratifying conclusion.

Whatever joy that might have been savored by the commander was left unstated. Perhaps there had been too many broken years to remember real pleasure, and how to respond to it. But his feelings could be seen in his eyes and in the little upturned crease at the corners of his mouth.

"You are a friend," Commander Freeman said at last.

John Pond smiled in return. He remembered a line from Emerson, one that he had not thought of in perhaps too many years: "The only way to have a friend is to be one."

CHAPTER 94

The *Guardian* begins to glide swiftly through the water as we head for the fire department's pier, where the fireboat is docked. It will take us a half hour or so to get in, and all the while we are caught up by this slowly moving image of San Francisco before us. It is a compelling city, and I cannot take my eyes from it. Chief Scott Peoples and I are sitting in the prow of the fireboat, taking the slap of every wave, and I feel the wind flat against my face.

A 24,000-gallons-per-minute pumping fireboat, the *Guardian* was put into service when the city bought it from Vancouver just after the Loma Prieta earthquake of 1989. During that quake, a 6.9 magnitude, it had been recognized how vital the work of another fireboat, the *Phoenix,* was in saving the Marina district, and since fireboats often go out of service for repair it was decided that the firefighters should have another fireboat always on call. The *Phoenix* had begun working for San Francisco in 1954, and has only a 9,600-GPM pumping capacity. Either fireboat can connect directly to the city's auxiliary water system in manifolds at different locations along the waterfront in case the regular water system fails, and they can also feed into the portable hydrant system an assistant chief named Frank Blackburn developed to bypass ruptured water mains in an emergency. The city and its fire department are very attentive to lessons, positive or negative, learned from the past.

The City of Oakland recently closed its fireboat as a cost-saving measure, but since they did not sell it they say that their idle vessel can always be reactivated in an emergency. But the pilots that were in the employ of the Oakland Fire Department have retired or left for other jobs, and it is unlikely that Oakland could man and run its fireboat within the narrow time window of an emergency. In spite of having one of the largest container ports in the United States, Oakland's authorities are willing to count on the *Guardian*'s being able to come across the bay for aid. Surely, the *Guardian* can play that role, contingent on just one thing: It would have to be available, and in every probability the *Guardian* would not be available in the event of another earthquake.

The wind is picking up, and the boat's sharp buffeting against the waves sends salt sprays up and over the prow. Oddly, this clean and washed feeling gives me a sense of adventure, perhaps the same excitement that brought the pioneers here to begin with. The novelty of being on this boat coupled with the excitement of being out on the water, as close to nature as one can get, etches itself into my memory. It is this immediate juxtaposition of big city life and being so close to the wilds of nature that is certainly one of the allures of San Francisco, for it is difficult to find a person in this city who does not routinely hike, camp, or sail. It is this background of love for the outdoors set against a stage of refinement and culture that has come to define the city and the people who live here.

"What is the best thing about San Francisco?" I ask Scott, perusing the shoreline from the Oakland Bay Bridge to the Ghirardelli sign almost to Fort Mason.

"You know," he answers, laughing. "I have thought about that many times. There are a thousand great things about this city, but I would say the real gem of San Francisco is its water system."

It is not such a surprising response, at least not from Scott. A water system is something that few of us ever really ponder, and it is only when we are forced to think about it we realize how wonderful and vital it is.

"That reminds me of Jupiter," I answer, knowing that Scott is as comfortable with science as with an eight-pound ax. "It is something we never think of, but it is a providential fact that, with its mighty gravitational size, Jupiter pulls into its body many meteorites and asteroids that might otherwise be headed straight for earth. There is no reason to ever think about this, but when we do we immediately see how everlastingly thankful we must be that Jupiter is there."

"Right," Scott says. "We can't imagine a world without Jupiter. And we can't imagine a world without water, either. But they sure realized what it was like in 1906."

And so this leads me to the question that is as inevitable as the rising sun when thinking about earthquakes in California, a question that should not be asked of a politician, the manager of a city agency, not even a fire commissioner—a question that should be asked only of someone who has spent a life and a career fighting fires, who knows what it is to send his firefighters into absolute peril, who knows that there are no second chances in doing things right in an emergency. Lives depend on what I want to know, which is: "What would happen today if another 8.0 earthquake shook San Francisco for forty-five seconds?"

We live in a country where there is very little accountability in business or in government. Just as we can be shocked that not one single businessman went to jail among the dozens who participated in graft schemes in 1906, we can be equally stunned by how few newspapers or public officials asked the congresspersons and U.S. senators who are members of the select intelligence committees to be accountable for the proven failures in intelligence leading up to 9/11. These are not idle comparisons when one considers that in both instances citizens believed that other issues were of more pressing importance, and so were willing to let charges of crime in 1906, and malfeasance in 2001, pass inconsequentially in order to progress into an orderly future. We do not demand resignations when cities or departments go into decline, or when our government fails to protect us. In America our leaders want only the glory of success with none of the attached responsibility when things go wrong.

I would not have much confidence in the response of political leaders to questions of preparedness. They would insist we are prepared, for to say otherwise suggests they have taken an eye off the ball, unless, conversely, they want people to think that an eye has been taken off the ball, as long as it is not their eye. But there are probably twenty fire chiefs in California whose answers to this question I would carefully attend to, and Scott is one of them.

"Well," he replied, "the United States Geological Survey tells us that there is a sixty-two percent probability of a major earthquake striking San Francisco in the next thirty years, but I think we should first look at 1906, the most serious conflagration in American history and what happened there. Essentially, the earthquake caused about two percent of the damage, and many lives, perhaps thousands, were lost in various collapses, including a policeman and two firemen. The numerous small fires that the shaking created were contained pretty early, and then because of a lack of manpower went out of control. This resulted in four major fires that burned 522 square city blocks, and at three of these fires—the Townsend Street–Rincon Hill fire, the Montgomery business district fire, and the Van Ness fire—there was Lieutenant Freeman of the navy leading the charge and assisting the SFFD. At the fourth fire, in the southern Mission district, the fire was stopped by a cistern at 19th and Shotwell, and with the help of a great many civilians at what we call the golden hydrant at 20th and Church.

"The San Francisco Fire Department then had 580 firemen and perhaps another hundred auxiliaries, all on duty, who were assisted by another hundred or so sailors and, at various times, civilian and other military personnel. In the whole year of 1906 the SFFD responded to fewer than 1,300 alarms of fire. But today, the city has just 340 firefighters on duty at any given time and responds to more than 144,680 calls a year. The SFFD has in place an important NERT [neighborhood emergency response team] program wherein it has trained and motivated a good number of willing and supportive civilians. It has been in operation for more than sixteen years and is a successful model for civilian-assisted emergency control. But who is to say how even those motivated and dedicated individuals will respond in an 8.0 earthquake to protect their own families, homes, and property?"

Scott's description makes it clear that the 1906 fire department had a distinct manpower advantage. Today, as many firefighters have told me, in a serious earthquake the bridges cannot be counted on for their approaches can easily be devastated, and so it will be difficult to supplement the manpower in the SFFD, even with an emergency recall plan (called Operation Return). Many of the 1,300 off-duty firemen will have to find ferry or helicopter transportation to San Francisco and then to their firehouses to pick up their emergency gear—no easy mission. Time and distance are the enemies of a fire chief in even the simplest of fires, and in a serious emergency both time and distance have to be integrated into any protection scenario. There can be a very serious delayed response by not only the incoming off-duty firefighters, but by every needed support system that may be called upon—like lifting equipment, cranes, and caravans of supply and equipment trucks, and, most especially, the responding fire companies from across the bay, down the peninsula, and more distant regions—*if* they are available and not tied up in their own emergency. Scott then suggested that in developing its plans San Francisco should consider itself an island and make certain that it has identified and filed every mobile water delivery system and earth-moving apparatus within the limits of the city.

Scott continued, "San Francisco's population in 1906 was 460,000. It is now 751,000 and is projected to be 925,000 by 2030. And the city's population swells to more than a million people during the typical workday, independent of the estimated 14 million tourists who visit us annually. In 1906, less than half of the geographic area of San Francisco was structurally developed, but today it is completely built out. In 1906, San Francisco had few high-rise buildings, but today has about

528 high-rise structures from eight to fifty-two stories in height, as well as 19,500 multiple-unit occupancies of three or more units.

"A significant issue for us is our urban population density, more dense in our streets and housing than any other city except New York—people mostly living in attached wooden structures that fill block after block. Most San Francisco streets have about twenty-four buildings attached side to side, but to an arriving fire company the street is seen as a fire hazard of one very large building with about thirty-thousand square feet of fire loading—a fire load that would require in most cities a whole fire department to control. And there are hundreds of such blocks in the city. This density simply adds to the challenge in protecting the lives of our citizens and our visitors. San Francisco has grown bigger in geography as well. Now, we have added the Presidio, Yerba Buena, and Treasure Island, and the airport—more protection responsibility for a strapped fire department."

As Scott speaks I remember a scientific analysis of earthquake and fire protection for San Francisco funded in 1987 by the insurance industry that was as close to Doomsday reportage as I have ever seen. It stated that the city, determined to rebuild quickly after the 1906 fire, overlooked or reduced the requirements to meet code, and so many of the buildings now standing in the city have wind loads, and floor and roof loads, that have been reduced by as much as 50 percent. The report predicted that in an earthquake similar to or worse than that of 1906, as many as 48,000 buildings would be destroyed in the ensuing fires—unless the city could dispatch 142 manned engine companies to control the fires that would develop. That is 100 engine companies more than the city actually has.

I think about the strapped fire departments—not only in San Francisco, but throughout the country. While it is true that some fire departments, because of political or union agreements, are reluctant to streamline, it should be remembered that there is a world of difference between streamlining and downsizing, which is what is called for today by many city managers. From coast to coast, the overwhelming pressure in city politics is to reduce costs, and that is brought about by a percentage reduction across the board in the budget of every city department. The problem for fire departments is that they have traditionally been well managed and efficient, and so mandatory reductions are inevitably damaging to the service they can provide. Like all generalizations, there are exceptions to this, but it is a safe observation.

San Francisco has instituted what it calls brownouts of fire companies in areas throughout the city, which have been explained as temporary solutions to budget shortfalls. These can be imposed on any fire department and already have in many of our biggest cities. Simply stated, the plan provides that a fire company goes out of service for a tour of twelve hours or twenty-four hours, and its firefighters are reassigned to other companies to fill manpower needs. In this way, the city circumnavigates the need to pay overtime to fill contracted manning requirements in individual fire companies. The companies are simply temporarily closed, and a closed company has no manpower requirement. About four to six companies each day are closed in San Francisco, and, in my mind, in a city where the threat of earthquake is ever present, a brownout will contribute to a critical reduced response time for a fire department and threatens to neutralize the department's readiness. Budget managers who inflict these solutions seem unwilling to recognize that fire departments are like insurance policies, where the premium relates directly to the amount of coverage. In almost all cases, the less a department spends, the less protection it can offer.

Response time can be crucial to saving lives, sometimes many lives if you consider that thousands of people may be facing death in a collapsing high-rise or in trains beneath the city on the BART (Bay Area Rapid Transit) system. San Francisco in 1906 never imagined high-rise buildings five times taller than the Spreckels Building or an underground mass transit system to Oakland, but today BART has a daily ridership of six hundred thousand. BART runs mostly underground and includes one of the longest underwater transit tubes in the world: six miles, including its approaches.

I asked Scott about this, and he answered, "You cannot forget that a major earthquake like an 8.0 will be exceptionally labor intensive, and in the first minute the fire department will undergo a large if not complete resource drain. SFFD will be required to respond to hundreds of immediate rescue needs, and not only in areas where liquefaction will cause collapse—there will be hundreds of general fire calls, hundreds of building collapse calls, and hundreds of emergency medical calls. There would be at least fifty significant fires, each attached or perilously close to another structure. These fires must be stopped, and the injured must be cared for. Every firefighter, police officer, EMT, and nurse in the city will be asked to make themselves available and will be challenged as never before.

"Let's look at Loma Prieta," Scott continued, "a 6.9 earthquake that lasted fifteen seconds, a good reference to use in projecting what may occur with a sustained shaking of 8.0 magnitude. In 1989, the Loma Prieta caused thirty-five fires with the most serious fire developing in the marina district. Structures in varying districts partly collapsed or suffered such damage that no one was ever allowed to go back into them, buildings that were ultimately red tagged and destroyed. The marina district fire provided everything we can expect in the future. It was like a theater production designed for study, complete with building collapses, a large fire, and the complications of a damaged water system with the SFFD fireboat coming to the rescue. Hundreds of citizens helped to lead water supply lines from the fireboat's bay location to the fire area. The firefighters displayed great heroism in placing their own lives secondary to those they rescued, working under explosive firefighting situations. It also placed incident commanders in difficult positions, forcing them to think creatively outside the box. Fire chiefs are used to a kind of war zone atmosphere during an uncontrolled fire, but at Loma Prieta there was also the pervasive collapse conditions with expectations of further collapse, and an initial confrontation and challenge when it was found that the water supply was damaged.

"On the other hand, not to diminish the terrible losses suffered by the marina district residents, a firefighter looking back at the marina fire could conclude that the fire occurred in one of the few favorable locations in San Francisco and under uncharacteristically good conditions. The marina district is bordered on three sides by firebreaks: the Marina Greens, the San Francisco Bay, and a six-lane divided roadway. When the water system failed, the marina district's proximity to the bay allowed the fireboat *Phoenix* to nose close to shore and substitute the water supply needed for firefighting. In addition, October 17, 1989 was an unusually warm day with no wind, inviting many people outdoors to the Marina Greens, saving many of them from the collapsed buildings. Every firefighter remembers how the smoke rose in one large column to the sky, almost as if bordered by a rule line, such was the absence of wind.

Finally, the earthquake occurred at approximately 5:04 P.M. during daylight hours. You couldn't buy another day like October 17, 1989 with the same favorable conditions. Still, there were three thousand calls between the earthquake and midnight, and the marina fire took the department to the maximum of its resources.

"We would be naïve not to anticipate the partial or total collapse of a number of buildings given our experience, and the possibility of an 8.0 sustained earthquake. Trains might derail, or worse, caused by the seismic thrust. Train and collapse problems will tax the SFFD's two heavy rescue units and the department's medium-size urban search and rescue [USAR] unit, besides demanding the efforts of our engine and truck companies. But these units will also be required at the large number of fires as the event progresses. The assistance of the few USAR units that exist in neighboring bay areas cannot be counted on because they might, and in all probability will, have their own emergencies. The mayor and the office of emergency services will have to act quickly in requesting the deployment of USAR units from across the country, with approval of the Federal Emergency Management Administration."

As we began to slowly back into the dock at the Harrison Street Pier, Pier 22½, and the quarters of Fireboat 1 and Engine 35, I began to think of how easy it is to take a fireboat from one spot to the other, but how on terra firma in a future significant earthquake the City of San Francisco could face insurmountable difficulty in moving apparatus, equipment, heavy rescue, and ambulance service through its streets. Great catastrophes often catch a population by surprise, and the ratiocination that has gone into making San Francisco an earthquake-prepared city is curious, for how can a mayor or a board of supervisors see that a department is brought to its limits by a 6.9 earthquake that shook for fifteen seconds and then demand that it close four or six fire companies each day to save money?

The State of California has mandated emergency preparedness, and it has done so by building an emergency structure that makes sense to the state. In the midst of an emergency, however, I worry that it would actually reduce the effectiveness of mitigation efforts.

In its plan, emergencies are broken down into three response levels: Level 1 is a street, or a building, or an incident disaster like a train derailment. Level 2 is for a local or neighborhood disaster, like a truck crash that empties a hazardous material into the air and requires a local evacuation. Level 3, the highest, is for a major disaster like Loma Prieta or an 8.0 earthquake. At Level 3, the city plan activates the creation of thirteen emergency response districts in the City of San Francisco, where each local battalion chief stops what he is doing and manages a rapid damage assessment for the city's emergency command center. This by definition diminishes the value of that battalion chief to the operations of the fire department while at the same time establishing his importance within the structure of the city's emergency operations plan.

At the emergency command center the mayor takes charge of the command section, where the police and fire chiefs and the chief health administrator report directly to him. Making the mayor the incident commander is a little like putting a ukulele player in charge of a symphony orchestra. It is not enough to know a little music. Indeed, in many American cities the formal emergency command structure pits the consideration of political officials, even with the best intentions, against the interests of public safety as might be seen by professional emergency leaders. In emergencies the person in charge must be someone who has made a life's career of protecting the lives and the security of citizens. The situation is not analogous to putting a politician in charge of a homeland security agency or a fire or police department, where the administration of a large coordinating effort takes place. It is small satisfaction to explain after a cataclysm that the presumption for incident command is that the mayor would listen to his professionals in an advise-and-consent way, the way a president at war is supposed to listen to the U.S. Congress. The fire commissioner in New York City has no voice in an actual firefighting or emergency effort for that is left to the chief of department. The City of San Francisco and the State of California might learn from that example.

The firefighters of the *Guardian* had placed a lunch of various sandwiches before us, and as we were eating I noticed a large sign hanging on the wall, one taken from the wheelhouse head of a now-retired fireboat. The *Guardian* was named in 1990 by a contest open to the children of San Francisco, and I cannot help contrasting this sign with the sign on the wall, which says DENNIS T. SULLIVAN, and identified a fireboat that had operated in San Francisco for fifty-four years.

The State of California had two fireboats built for the Port of San

Francisco after the great fire. The boats were modeled after those of New York's harbor. One was named for David Scannel, a colorful fire chief from the 1870s who is also the namesake of one of SFFD's most important medals of honor and that was stationed where the *Guardian* is now. The other was the *Sullivan,* named after the hero chief of the turn of the last century, and it had been stationed at Pier 31. Eleven firefighters manned each fireboat, and they served the city until 1954.

I cannot help reflecting that after all the firefighters did for the City of San Francisco during the great fire of 1906, not only is there not a monument to their courage and unquestioned dedication during that tragic time but the City of San Francisco, in naming its most recent fireboat, replaced a heroic individual with a concept. This decision is consistent with notions of contemporary art that replace the human with the conceptual. But the firefighters sitting around the lunch table with Scott and me know that when a building is on fire, concepts, even those found in words like *courage, commitment, hero,* and *guardian* won't put out the fires. It is people who will.

When I mentioned this to the firefighters, they laughed. They have no regrets. They love the city almost as much as they love their job. And whenever the city decides to honor its firefighters for the duty, deaths, and dedication they contributed to bringing the fire of 1906 under control that will be fine with them.

CHAPTER 95

On the Saturday following the death of Frederick Freeman, Commander, United States Navy (ret.), his loyal and admiring benefactor, John Pond, Commander, United States Navy (ret.), was the sole passenger on a friend's boat. When the vessel reached some distance from the line of wharfs at the north end of San Francisco, Pond asked his friend to stop the boat and turn it toward San Francisco. As this was being done Pond lifted a container from a canvas bag. The city seemed to sparkle in its undulating rhythms before him, the thousands of buildings running up and down the hills standing tall and straight as if in salute. He opened the container, a vaselike porcelain pot, placing its lid in his lap. He then deliberately painted in his mind's eye that picture of Frisky Freeman he had remembered so often. There he was, leading the sailors and the firemen into the smoke-clouded street where Jackson meets Montgomery, his finger pointing in the direction of the flames, calling out his familiar, "Okay, men, let's sock it to 'em!"

Pond then held the container and slowly poured its contents out over the ancient waters of San Francisco Bay. The ashes of Frisky Freeman then melded and became one with the busy harbor he had worked so hard to save. Pond smiled, relieved. Lieutenant Freeman had finally returned to San Francisco in the proper way—with honor.

AUTHOR'S NOTE

This is not an academic work, but a historical narrative, and I have not footnoted the facts so that the reader would not be diffused in the reading. However, all information contained in the book may be relied upon as historically accurate.

U.S. ARMY TROOPS ASSIGNED TO THE SAN FRANCISCO TRAGEDY

Fort Mason, San Francisco
Companies D & D of the Engineering Corps
Captain M. L. Walker, commanding
Arrived on April 18

Presidio, San Francisco
10th, 29th, 38th, 66th, 67th, 70th, and 105th Companies from the Coast Artillery
Troops I and K of the 14th Cavalry
1st, 9th, and 24th Field Batteries
Colonel Charles Morris, commanding
Arrived on April 18

Fort Miley, San Francisco
25th and 64th Companies of the Coast Artillery
Major C. H. Hunter, commanding
Arrived on April 18

Fort McDowell (Angel Island)
22nd Infantry Regiment
Colonel Alfred Reynolds, commanding
(Note: Most were at a firing range near Point Bolinas at the time of the earthquake.)
Arrived on April 18

Presidio, Monterey
20th Infantry Regiment (recently arrived from fighting in the Philippines)
Colonel Marion P. Maus, commanding
Arrived on April 19

Vancouver Barracks (Vancouver, Washington)
14th Infantry Regiment, ten companies (including headquarters staff and regimental band)
Arrived on April 22nd
14th Infantry Regiment, 17th and 19th Batteries of Field Artillery
Arrived on April 23

BIBLIOGRAPHY

BOOKS

Aczel, Amir D., *The Riddle of the Compass,* San Diego: Harcourt, 2001.

Adler, Jacob, *Claus Spreckels: The Sugar King in Hawaii,* Honolulu: University of Hawaii Press, 1966.

Alexander and Heig, *San Francisco: Building the Dream City,* San Francisco: Scottwall Associates, 2002.

Altrocchi, Julia C., *The Spectacular San Franciscans,* New York: Dutton, 1949.

Adjuster, The (an insurance journal), San Francisco: June 1906.

Ashton, John F., *In Six Days,* Green Forest, AK: Master Books, 2000.

Bain, David H., *Sitting in Darkness: Americans in the Philippines,* Boston: Houghton, Mifflin, 1984.

Bancroft, Hubert, *Some Cities and San Francisco,* New York: Bancroft Co., 1907.

Bawlf, Samuel, *The Secret Voyage of Sir Francis Drake,* London: Douglas & McIntyre, 2003.

Bell, F.G., *Geologic Hazards,* London and New York: E&FN, 1999.

Birmingham, Stephen, *California Rich,* New York: Simon & Schuster, 1980.

Block, Eugene B., *The Immortal San Franciscans,* San Francisco: Chronicle Books, 1971.

Bohemian Club, *Annals of the Bohemian Club,* San Francisco: privately printed, 1990.

Bonnet, Theodore F., *The Regenerators,* San Francisco: Pacific Printing Co., 1911.

Brammer, Alex, *Victorian Classics of San Francisco,* Sausalito, CA: Windgate Press, 1987.

Bronson, William, *The Earth Shook, the Sky Burned,* San Francisco: Chronicle Books, 1959.

Brown, Kenneth A., *Cycles of Rock and Water,* New York: HarperCollinsWest, 1992.

Bryant, Edward, *Natural Hazards,* Cambridge and New York: Cambridge University Press, 1991.

Coffman, William M., *American in the Rough,* New York: Simon & Schuster, 1955.

Conlon, John J., *Eyewitness Account,* Museum of the City of San Francisco, 1982.

Courland, Robert, *The Old North Waterfront,* San Francisco: Ron Kaufman Companies, 2004.

Davis, Kenneth C., *Don't Know Much About History,* New York: HarperCollins, 2003.

Dickson, Samuel, *San Francisco Kaleidoscope,* Stanford: Stanford University Press, 1949.

Ditzel, Paul, *Fire Engines, Fire Fighters,* New York: Crown, 1976.

Dodge, Richard V., *Rails of the Silver Gate,* San Marino, CA: Pacific Railway Journal, 1960.

Dolan, Edward F., Jr., *Disaster, 1906,* New York: Julian Messner, 1959.

Dowling, Patrick J., *Irish Californians,* San Francisco: Scottwall Associates, 1998.

Dupuy, Trevor, Curt Johnson, and David L. Bongard, *The Harper Encyclopedia of Military Biography,* New York: HarperCollins, 1992.

Edwords, Clarence, *Bohemian San Francisco,* San Francisco: Paul Elder & Co., 1914.

Fox, Stephen, *John Muir and His Legacy,* Boston: Little, Brown, 1981.

Fradkin, Philip L., *Magnitude 8,* New York: A John Macrae Book, Henry Holt, 1998.

Funston, General Frederick, *Memories of Two Wars,* New York: Charles Scribner's Sons, 1911. A fascinating account of Funston's participation in the Cuban rebellion and his experiences as a colonel, then general, commanding American troops in the Philippines. He judiciously passes over the more gruesome measures he took against the Philippine insurgents.

Genthe, Arnold, *As I Remember,* New York: Regnal & Hitchcock, 1934.

Gentry, Curt, *The Dolphin Guide to San Francisco,* Garden City, NY: Dolphin Books, 1962.

Graham, Otis Jr., *Unguarded Gates: A History of America's Immigration Crises,* New York: Rowman & Littlefield, 2004.

Hansen, Gladys, and Emmet Condon, *Denial of Disaster,* San Francisco: Cameron and Company, 1989.

Harriman, E. Roland, *I Reminisce,* Garden City, NY: Doubleday, 1975.

Haywood, Charles F., *General Alarm,* New York: Dodd Mead, 1967.

Heitz, William, *Wine Country,* Santa Barbara, CA: Capra Press, 1990.

Issel, William, and Robert Cherny, *San Francisco 1865–1932,* Berkeley: University of California Press, 1986.

Jones, Idwal, *Arc of Empire,* Garden City, NY: Doubleday, 1951.

Kahn, Edgar, *Cable Car Days in San Francisco,* San Francisco: San Francisco Public Library Publications, 1976.

Kendrick, T.D., *The Lisbon Earthquake,* New York: Lippincott, 1955.

Kennedy, John C., *The Great Earthquake and Fire,* New York: William Morrow, 1963.

Kenoger, Natlee, *The Firehorses of San Francisco,* Los Angeles: Westernlore Press, 1970.

Kirk, Anthony, *A Flier in Oil,* San Francisco: California Historical Society, 2000.

Kurzman, Dan, *Disaster,* New York: William Morrow, 2001.

McClain, Charles, *In Search of Equality,* Berkeley: University of California Press, 1994.

McCullough, David, *The Path Between the Seas,* New York: Simon & Schuster, 2004.

Martin, Mildred, *Chinatown's Angry Angel: The Story of Donaldina Cameron,* Palo Alto, CA: Pacific Books, 1977.

Moffat, Frances, *Dancing on the Brink of the World,* New York: Putnam, 1977.

Muscatine, Doris, *Old San Francisco,* New York: Putnam, 1975.

Obenzinger, Hilton, *Cannibal Eliot and the Lost Histories of San Francisco,* San Francisco: Mercury House, 1993.

Older, Fremont, *My Own Story*, San Francisco: The Call Publishing Co., 1919.

Olmsted, Roger, and T. H. Watkins, *Here Today—San Francisco's Architectural Heritage,* San Francisco: Chronicle Books, 1968.

Oppel, Frank, *Tales of California,* Secacus, NJ: Castle Books, 1989.

Rand McNally Co., *San Francisco and Oakland Guide,* New York, 1903.

Reisner, Marc, *A Dangerous Place,* New York: Pantheon Books, 2003.

Riis, Jacob, *How the Other Half Lives,* New York: Charles Scribner's Sons, 1890.

Ritchie, David, *Superquake,* New York: Crown, 1988.

Shumat, Albert, *Rincon Hill and South Park,* Sausalito, CA: Wingate Press, 1988.

Smith, Dennis, *Dennis Smith's History of Firefighting in America,* New York: Dial Press, 1978.

Steele, Rufus, *The City That Is,* San Francisco: A. M. Robertson, 1909.

Steffens, Lincoln, "The Mote and the Beam," *American Magazine,* November 1907.

Stetson, James B., *Narrative of My Experiences in the Earthquake and Fire of 1906,* Palo Alto, CA: Lewis Osborne Books, 1969.

Thomas, Gordon, and Max Morgan Witts, *The San Francisco Earthquake,* New York: Stein and Day, 1971.

Thomas, Lately, *A Debonair Scoundrel,* New York: Holt, Rinehart, and Winston, 1962.
Tributsch, Helmut, *When the Snakes Awake: Animals and Earthquake Prediction,* Cambridge, MA: MIT Press, 1982.
Tyler, Sydney, *San Francisco's Great Disaster,* Philadelphia: P.W. Ziegler Co., 1906.
Victor, G., and Brett Nee de Bary, *Longtime Californ',* Boston: Houghton, Mifflin, 1972.
Walsh, James, and Timothy O'Keefe, *Legacy of a Native Son,* San Francisco: Forbes Mill Press, Montalvo Association, 1993.
Watkins, T.H., *California: An Illustrated History,* New York: Weathervane Books, 1983.
Wegener, Alfred, *The Origins of Continents and Oceans,* New York: Dover Publications, 1966.
Wells, Evelyn, *Fremont Older,* New York: D. Appleton-Century Co., 1936.
Whitney, James, *The Chinese and the Chinese Question,* San Francisco: R&E Research, 1888.

ARTICLES, REPORTS, AND LETTERS

Account of Marine Daniel Morgan, of the USS *Marblehead* during the disaster, quoted in *The Socialist Voice,* Oakland, April 28, 1906.
"Army and Navy News," *The Argonaut,* March 24, 1906, San Francisco Public Library, Main Branch, The Koshland San Francisco History Room.
Army records from the National Archives and Records Administration, Washington, D.C.
"A Bit of Forgotten History, the Navy at San Francisco, April 18–23," 1906 by Commander John E. Pond (Retired), U.S.N. (unpublished) 1931, the Pond Family Collection.
"Beneath the Surface," *New York Times,* July 2, 1907.
Diary entries for April 18, 1906, of James D. Phelan, Phelan Family Collection.
"A History of the Twenty-second United States Infantry" by Captain W. H. Wassell, 22nd Infantry, Manila, Philippines, 1904. Because of the author's obvious bias and the fact that the previous thirty years of his unit's otherwise distinguished service history involved controversial military actions (the Sioux uprisings, the Coeur d'Alene mine strike, and the Philippine Insurrection), this account is as much cloaked as it is informative. San Francisco: Presidio Archives, National Golden Gate Recreation Area, National Park Service.
"How the Army Worked to Save San Francisco" by Frederick Funston, U.S.A. Brigadier General, *Cosmopolitan* magazine, July 1906, the Hansen Collection.
"How the Navy Saved the Waterfront," *The Argonaut,* August 25, 1906, San Francisco Public Library, Main Branch, The Koshland San Francisco History Room.
Mare Island Shipyard Logbook, June 1, 1905–September 30, 1906, Record Group 181, #591 (MINSY), National Archives and Records Administration (Northern California Branch), San Bruno, California.
Navy records from the National Archives and Records Administration, Washington, D.C.
Interview with George F. Williams, *New York Sun,* April 19, 1906, the Hansen Collection.
Science Times Supplement, *New York Times,* September 21, 2004.
Oakland Tribune, April 30, 1906 (partial list of buildings saved by firefighting efforts, eighteen of the twenty-two mentioned were through Freeman's efforts, and those of elements of the SFFD and army assisting him).
"Observation of the San Francisco Earthquake," Joseph H. Harper, *Montana Society of Engineers Journal,* 1908.

Official report by Captain Le Vert Coleman, May 2, 1906, the Hansen Collection. Discovered by David Fowler at the National Archives and Records Administration.

Personal letter written by PFC De Loa R. Kaff, April 22, 1906, the Hansen Collection.

Presidial Weekly Clarion, Friday, April 27, 1906.

Report by Captain D.E. Aultman. Reference number: 2844, San Francisco: Presidio Archives, National Golden Gate Recreation Area, National Park Service.

Report by Lieutenant Frederick N. Freeman, U.S. Navy, Commanding USTBD *Perry,* Mare Island, California, April 30, 1906. Reference code: 175-F, the Hansen Collection. Original copy at the National Archives and Records Administration (Northern California Branch), San Bruno, California.

Report by Warrant Officer O. D. Miller, U.S.A., executive assistant and chief clerk, Quartermaster Section and Army Transport Service, Fort Mason, California, April 18, 1906, then serving as civil service clerk, Quartermaster Corps, on duty as record clerk, office Depot-Quartermaster, 35 New Montgomery Street. Forwarded by Lieutenant Colonel Benjamin H. L. Williams, Fort Mason, California, April 8, 1931, re: "Participation in Army intervention in San Francisco Fire and Earthquake April 18, 1906, San Francisco: Presidio Archives, National Golden Gate Recreation Area, National Park Service.

Scrapbook of Major Carroll A. Devol, Box 1 of 2 (GOGA 32485) of Devol Collection (GOGA-2048). San Francisco: Presidio Archives, National Golden Gate Recreation Area, National Park Service.

Special report of Major General Adolphus Greely, U.S.A., commanding the Pacific Division, on the relief operations conducted by the military authorities of the United States at San Francisco and other points, with accompanying documents, Washington, D.C.: Government Printing Office, 1906. (Note: About 95 percent of Greely's extensive report covers relief operations after the fire and the results of his investigation into charges that army personnel stole property during the disaster. The relief operations were brilliantly handled by Major Carroll Devol and helped hundreds of thousands of distressed refugees. Greely cites criminal investigations of certain army personnel, but these involved relatively petty charges. There were no investigations into the summary executions of individuals by soldiers of the army since Greely followed Funston's lead and maintained that no such killings were committed by his men.) San Francisco Public Library, Main Branch, The Koshland San Francisco History Room. Funston's original typewritten copy is at the Presidio Archives, National Golden Gate Recreation Area, National Park Service, San Francisco, but the first and last pages have been lost sometime over the last ninety-nine years.

"Fighting the Fire," *The Argonaut,* June 16, 1906, the San Francisco Public Library, Main Branch, The Koshland San Francisco History Room.

The Bulletin, San Francisco, April 20, 1906.

"The California National Guard in the San Francisco Earthquake and Fire of 1906," James J. Hudson, *California History Quarterly,* Summer 1976.

The Call–Bulletin, special issue. April 19, 1906.

"The Great Fire of 1906, Adventures of Grand Hotel Guests. Mr. James W. Bryant's Experiences at the Palace," Chapter XXXII, *The Argonaut,* November 27, 1926, San Francisco Public Library, Main Branch, The Koshland San Francisco History Room.

"The Great Fire of 1906," Chapter XIV, *The Argonaut,* July 24, 1927, San Francisco Public Library, Main Branch, The Koshland San Francisco History Room.

"The Great Fire of 1906, Friction Between the Citizens Committee and the Militia—How It Happened," Chapter LX, *The Argonaut,* June 11, 1927, San Francisco Public Library, Main Branch, The Koshland San Francisco History Room.

"The Great Fire of 1906, How Mayor Schmitz Rose to the Emergency of the Disaster," Chapter XXXIX, *The Argonaut,* January 15, 1907, San Francisco Public Library, Main Branch, The Koshland San Francisco History Room. (Note: Includes accounts by John T. Williams and Myrtile Cerf on Schmitz's activities in the first hours after the earthquake.)

"The Great Fire of 1906, How the Fire Companies Were Scattered Inefficiently When the Fire Started," Chapter LXVII, *The Argonaut,* June 30, 1927, San Francisco Public Library, Main Branch, The Koshland San Francisco History Room.

"The Great Fire of 1906, How the Fires Spread and Joined—Dynamiting Starts Blaze Afresh," Chapter LXIX, *The Argonaut,* August 13, 1927, San Francisco Public Library, Main Branch, The Koshland San Francisco History Room.

"The Great Fire of 1906, The Fires in the Produce District—Experiences of a Civilian with a Policeman's Badge," Chapter V, *The Argonaut,* May 29, 1926, San Francisco Public Library, Main Branch, The Koshland San Francisco History Room.

"The Great Fire of 1906, Use of Dynamite to Stop Spread of Fire—Military Objections to Volunteer Firemen," Chapter LXVIII, *The Argonaut,* June 30, 1927, San Francisco Public Library, Main Branch, The Koshland San Francisco History Room.

"The Great Fire of 1906, Refugee as Steeple-Climber—How Montgomery Block Was Saved—Hotaling Whiskey," Chapter XIII, *The Argonaut,* July 17, 1926, San Francisco Public Library, Main Branch, The Koshland San Francisco History Room.

"The Human Drama of San Francisco," Herman Whitaker, *Harper's Weekly,* May 1906.

"Complete Destruction Is Conceded by Funston" and "Shock Brought Terror," *Topeka Daily Capital,* Friday, April 20, 1906.

Transcription from photocopy of handwritten manuscript, letter from soldier (signature indecipherable), dated April 23, 1906, the Hansen Collection.

Report on the Underground Water Supply System, M. M. O'Shoughnessy, City of San Francisco, 1913.

Reports on an Auxiliary Water Supply System, Marsden Manson, Britton & Rey, San Francisco, 1908

RG 125 navy examining boards: Frederick Freeman.

RG 45 U.S. Navy subject file, 1775–1910, file OO Box 464 (11W4 16/3/3).

RG 80 secretary of the navy correspondence, 1897–1915, file 21780 Box 778 (photos) 11W3 20/6/6.

RG 94, AGO doc file, 1121191 Box 4457–4459 (8W3 15/26/5).

RG 107, secretary of war general correspondence, 1890–1913, "San Francisco earthquake," Box 87 (16W3 8/29/2).

RG 393 Part I, entry 521 Department of California, correspondence re San Francisco fire.

RG 77, Index to Engineers Doc file, 25 index cards "San Francisco, earthquake and fire" Box 261 (8W4 5-5-4).

RG 92 Index to Quartermaster's Doc File (five boxes for San Francisco)

Note: Regarding the above boxes (both army and navy), some records in one file appeared to be mixed up with those of another. For example, most but not all War Department cables and telegrams about the disaster were in the files from the Office of the Adjutant (Military Secretary's Office) RG 94, 1121370–1121191, Box 4457 or RG 107, secretary of war's correspondence, 1890–1913, Box 87 (16W3 8/29/2).

Files regarding legal suits seeking compensation for stolen or destroyed goods were generally, but not always, in RG 94, 1121 191–End, Boxes 4458 and 4459. Records regarding actions taken by the U.S. Navy and Lieutenant Frederick N. Freeman were generally but not always in RG 45, U.S. Navy Subject File, 1775–1910, File OO Box 464 (11W4 16/3/3) or RG 80 secretary of the navy correspondence, 1897–1915, file 21780 Box 778 (photos) 11W3 20/6/6. These misplacements could have occurred earlier, when another researcher examined the files and mistakenly put them back in the wrong folder.

INDEX